GRAPH THEORY

annals of discrete mathematics

General Editor
Peter L. HAMMER, University of Waterloo, Ont., Canada

Advisory Editors
C. BERGE, Université de Paris, France
M.A. HARRISON, University of California, Berkeley, CA, U.S.A.
V. KLEE, University of Washington, Seattle, WA, U.S.A.
J.H. VAN LINT, California Institute of Technology, Pasadena, CA, U.S.A.
G.-C. ROTA, Massachusetts Institute of Technology, Cambridge, MA, U.S.A.

NORTH-HOLLAND PUBLISHING COMPANY – AMSTERDAM • NEW YORK • OXFORD

NORTH-HOLLAND
MATHEMATICS STUDIES　　　　　　　　　　　　　62

Annals of Discrete Mathematics (13)

General Editor: Peter L. Hammer

University of Waterloo, Ont., Canada

Graph Theory

Proceedings of the Conference
on Graph Theory, Cambridge

Editor:

Béla BOLLOBÁS

*Department of Pure Mathematics
and Mathematical Statistics
University of Cambridge
Cambridge CB2 1SB, England*

1982

NORTH-HOLLAND PUBLISHING COMPANY – AMSTERDAM • NEW YORK • OXFORD

© *North-Holland Publishing Company, 1982*

All rights reserved. No part of this publication may be reproduced, stored in a retrieval system or transmitted, in any other form or by any means, electronic, mechanical, photocopying, recording or otherwise, without the prior permission of the copyright owner.

ISBN: 0 444 86449 0

Publishers:
NORTH-HOLLAND PUBLISHING COMPANY
AMSTERDAM • NEW YORK • OXFORD

Sole distributors for the U.S.A. and Canada:
ELSEVIER SCIENCE PUBLISHING COMPANY, INC.
52 VANDERBILT AVENUE, NEW YORK, NY 10017

Library of Congress Cataloging in Publication Data
Main entry under title:

Graph theory.

(Annals of discrete mathematics ; 13) (North-Holland mathematics studies ; 62)
Papers presented at the Cambridge Graph Theory Conference, held at Trinity College 11-13 Mar. 1981.
1. Graph theory--Congresses. I. Bollobás, Béla. II. Cambridge Graph Theory Conference (1981 : Trinity College, University of Cambridge) III. Series. IV. Series: North-Holland mathematics studies ; 62.
QA166.G718 1982   511'.5   82-8098
ISBN 0-444-86449-0   AACR2

PRINTED IN THE NETHERLANDS

FOREWORD

The Cambridge Graph Theory Conference, held at Trinity College from 11 to 13 March 1981, brought together top ranking workers from diverse areas of the subject. The papers presented were by invitation only. This volume contains most of the contributions, suitably refereed and revised.

For many years now, graph theory has been developing at a great pace and in many directions. In order to emphasize the variety of questions and to preserve the freshness of research, the theme of the meeting was not restricted. Consequently, the papers in this volume deal with many aspects of graph theory, including colouring, connectivity, cycles, Ramsey theory, random graphs, flows, simplicial decompositions and directed graphs. A number of other papers are concerned with related areas, including hypergraphs, designs, algorithms, games and social models. This wealth of topics should enhance the attractiveness of the volume.

It is a pleasure to thank Mrs. J.E. Scutt and Mrs. B. Sharples for retyping most of the papers so quickly and carefully, and to acknowledge the financial assistance of the Department of Pure Mathematics and Mathematical Statistics of the University of Cambridge. Above all, warm thanks are due to the participants for the exciting lectures and lively discussions at the meeting and for the many excellent papers in this volume.

Béla Bollobás

25th February 1982
Baton Rouge

TABLE OF CONTENTS

Foreword	v
J. AKIYAMA, K. ANDO and H. MIZUNO, Characterizations and classifications of biconnected graphs	1
R.A. BARI, Line graphs and their chromatic polynomials	15
J.C. BERMOND, C. DELORME and G. FARHI, Large graphs with given degree and diameter III	23
B. BOLLOBÁS, Distinguishing vertices of random graphs	33
B. BOLLOBÁS and F. HARARY, The trail number of a graph	51
J.H. CONWAY and M.J.T. GUY, Message graphs	61
D.E. DAYKIN and P. FRANKL, Sets of graph colourings	65
P. DUCHET and H. MEYNIEL, On Hadwiger's number and the stability number	71
H. DE FRAYSSEIX and P. ROSENSTIEHL, A depth-first-search characterization of planarity	75
J.L. GROSS, Graph-theoretical model of social organization	81
R. HÄGGKVIST, Odd cycles of specified length in non-bipartite graphs	89
R. HALIN, Simplicial decompositions: Some new aspects and applications	101
F. HARARY, Achievement and avoidance games for graphs	111

A.J.W. HILTON, Embedding incomplete latin rectangles 121

A.J.W. HILTON and C.A. RODGER, Edge-colouring regular
bipartite graphs 139

A.J. MANSFIELD and D.J.A. WELSH, Some colouring problems
and their complexity 159

A. PAPAIOANNOU, A Hamiltonian game 171

J. SHEEHAN, Finite Ramsey theory and strongly regular graphs 179

R. TINDELL, The connectivities of a graph and its complement 191

CHARACTERIZATIONS AND CLASSIFICATIONS OF BICONNECTED GRAPHS

J. AKIYAMA, K. ANDO and H. MIZUNO

Nippon Ika University, Kawasaki, Japan
and
University of Electrocommunications, Tokyo, Japan

Dedicated to Frank Harary for his 60th birthday

A graph G is said to be *biconnected* if both G and its complement $\bar{G}$ are connected. In this paper we mainly deal with biconnected graphs and classify a set $\mathcal{G}$ of biconnected graphs into several classes in terms of the number of cutvertices (or endvertices) of G and $\bar{G}$. Furthermore we give structural characterizations of these classes, that is, characterizations of graphs G such that G has m cutvertices (endvertices) and $\bar{G}$ has n cutvertices (endvertices) respectively, where m and n are given, $1 \leq m \leq 2$ and $1 \leq n \leq 2$.

§1. INTRODUCTION

A graph G is *biconnected* if both G and its complement $\bar{G}$ are connected. In this paper we are mainly concerned with biconnected graphs. We denote by $\mathcal{G}$ the set of all biconnected graphs and by $\mathcal{G}_p$ the set of all biconnected graphs of order p. We use the notations and terminology of [1] or [2]. The characterization of biconnected graphs is well known, which is stated as follows:

Theorem A. [1, p. 26] A graph G of order $p(p \geq 2)$ is biconnected if and only if neither G nor $\bar{G}$ contains a complete bigraph $K(m,n)$ as a spanning subgraph for some m and n with $m + n = p$.

We denote by $c(G)$, $e(G)$ the number of cutvertices and endvertices of a connected graph G, respectively. We say the *cutvertex-type* of $G \in \mathcal{G}$ is (m,n) if $\{c(G), c(\bar{G})\} = \{m,n\}$. Similarly the *endvertex-type* of $G \in \mathcal{G}$ is (m,n) if $\{e(G), e(\bar{G})\} = \{m,n\}$. Note that the cutvertex-type (the endvertex-type) of a graph $G \in \mathcal{G}$ is the same as one of its complement $\bar{G}$ from the definition. The cutvertex-type and the endvertex-type for all graphs of order $p \leq 5$, as presented in Table 1.

type graph	cutvertex-type	endvertex-type
	(2,2)	(2,2)
	(2,2)	(2,2)
	(0,0)	(0,0)
	(2,1)	(1,1)
	(2,1)	(3,1)
	(3,0)	(2,0)

Table 1. The cutvertex-type and the endvertex-type for graphs ($\in \mathcal{G}$) of order p (≤ 5).

We now present a list of symbols which we use in this paper:
$C(m,n) = \{G \in \mathcal{G} \mid G$ is a graph with cutvertex-type $(m,n)\}$,
$E(m,n) = \{G \in \mathcal{G} \mid G$ is a graph with endvertex-type $(m,n)\}$,
$C_p(m,n) = C(m,n) \cap \mathcal{G}_p$,
$E_p(m,n) = E(m,n) \cap \mathcal{G}_p$,
$C(G) = \{v \in V(G) \mid v$ is a cutvertex of $G\}$,
$End(G) = \{v \in V(G) \mid v$ is an endvertex of $G\}$,
$E = \bigcup_{m,n \geq 1} E(m,n)$, that is E is a set of all biconnected graphs G such that both G and $\bar{G}$ have at least one endvertex.

The following two well known results are required to prove the lemmas we will mention later.

<u>Proposition 1</u>. Every nontrivial graph G of order p contains at least two vertices that are not cutvertices;

$$c(G) \leq p - 2 \text{ for every nontrivial graph } G. \qquad \square$$

<u>Proposition 2</u>. (Harary [3]) If v is a cutvertex of a biconnected graph G, then v is not a cutvertex of $\bar{G}$;

$$C(G) \cap C(\bar{G}) = \phi \text{ for every graph } G \in \mathcal{G}. \qquad \square$$

We now present two fundamental lemmas.

Lemma 1. Let G be a connected graph and v be any vertex not in $C(G)$. Then

(i) $c(G - v) \geq c(G) - 1$,
(ii) and if the equality holds, then v is an endvertex of G.

Proof. Let $C(G) = \{u_1, u_2, \ldots, u_n\}$ be the collection of cutvertices of G, where $n = c(G)$. Assume that $G - u_i = \bigcup_{k=1}^{m_i} G_k^{(i)}$, where $G_k^{(i)}$ ($1 \leq k \leq m_i$) are components of $G - u_i$. Without loss of generality we may assume that $G_1^{(i)}$ contains v. We divide our proof into two cases depending on the degree of v.

Case 1. $\deg_G v \geq 2$. In this case, the degree of v in $G_1^{(i)}$ is at least 1, and so $G_1^{(i)} - v$ is not a null graph. Thus, for every $i (1 \leq i \leq n)$, $G - u_i - v = (G - v) - u_i = (G_1^{(i)} - v) \cup G_2^{(i)} \cup \ldots$
$\ldots \cup G_{m_i}^{(i)}$ holds. This fact implies that u_i is also a cutvertex of $G - v$. Therefore we obtain the inequality

$$c(G - v) \geq c(G). \qquad (1)$$

Case 2. $\deg_G v = 1$. In this case there are two possibilities depending on whether v is adjacent to a cutvertex of $G - v$ or not.

$$c(G - v) \geq c(G) \qquad (2)$$

or

$$c(G - v) \geq c(G) - 1. \qquad (3)$$

By (1), (2) and (3) we complete the proof. □

Remark: Lemma 1 gives the lower bound of $c(G - v)$ in terms of $c(G)$ for a connected graph G and a non-cutvertex v of G, however it is impossible to estimate the nontrivial upper bound of $c(G - v)$ in terms of $c(G)$ as seen in the example $G = P_n + K_1$. Furthermore, note that the converse of statement (ii) of Lemma 1 does not hold when the endvertex v of G is adjacent to a cutvertex of $G - v$, as shown by the graph in Figure 1.

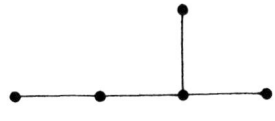

Figure 1

It is easy to prove the following results corresponding the previous two propositions.

Proposition 3. For any biconnected graph G,
$$e(G) \leq p - 2.$$

Proof. For any connected graph H of order p, the inequality $e(H) \leq p(H) - 1$ holds, and if the equality holds, namely $e(H) = p(H) - 1$, then H must be a star. However every star is not biconnected, thus we obtain
$$e(G) \leq p - 2 \quad \text{for any} \quad G \in \mathcal{G}. \qquad \square$$

Proposition 4. If v is an endvertex of a biconnected graph G, then v is not an endvertex of $\bar{G}$;
$$\text{End}(G) \cap \text{End}(\bar{G}) = \phi \quad \text{for every graph} \quad G \in \mathcal{G}. \qquad \square$$

Lemma 2. Let $G \in \mathcal{G}_p$, and u, v be any vertices with $\deg_G u = 1$, $\deg_G v = p - 2$, that is, $u \in \text{End}(G)$ and $v \in \text{End}(\bar{G})$. Then, G contains a path P_4 as an induced subgraph, which contains both u and v.

Proof. We may assume that $uv \in E(G)$, otherwise we may consider uv in $\bar{G}$ instead of G. Since $\deg_G v = p - 2$, there is exactly one vertex w which is not adjacent to v in G. Since $\deg_G u = 1$ and $uv \in E(G)$ by the assumption, u is not adjacent w in G. Since G is connected, w is adjacent to some vertex x other than u or v. Since $\deg_G v = p - 2$ and $\deg_G u = 1$, we see that v is adjacent to x, and u is not adjacent to x. (See Figure 2.)

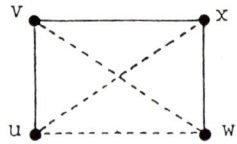

Figure 2.

Therefore the subgraph induced by $\{u, v, w, x\}$ in G (and also in $\bar{G}$) is P_4 containing both u and v as required. $\qquad \square$

If G and P_n are graphs with the property that the identification of any vertex of G with an arbitrary endvertex of H results in a unique graph (up to isomorphism), then we write $G \cdot P_n$ for this graph.

We shall show the existence theorem for graphs with certain cut-vertex-types or endvertex-types.

<u>Theorem 1</u>. For all positive integers $p \geq 6$ and $m(0 \leq m \leq p - 2)$,

(i) $C_p(m,0) \neq \emptyset$,

(ii) $E_p(m,0) \neq \emptyset$

and

(iii) $E_p(m,1) \neq \emptyset$.

<u>Proof</u>. To prove this, it is sufficient to exhibit graphs $G \in \mathcal{G}$ belonging to the various sets (see Figure 3):

(i) C_p $(m = 0)$,

           $C_{p-m} \cdot P_{m+1}$ $(1 \leq m \leq p - 1)$,

           P_p $(m = p - 2)$.

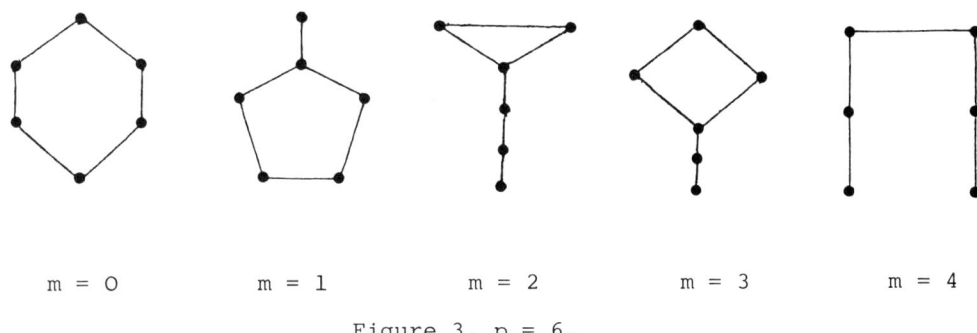

m = 0   m = 1   m = 2   m = 3   m = 4

Figure 3. $p = 6$.

(ii) Let $T_m = T(m/2, m/2)$ be a double star and define a graph T'_m as the graph obtained from T_m by inserting $p - m - 2$ new vertices of degree 2 into the edge joining two nonendvertices of T_m (see Figure 4). Then the graph T'_m is in $E(m,0)$ for $m(0 \leq m \leq p - 2)$.

Figure 4. $p = 8$, $m = 5$, when $m = 0$, it coincides with the case (ii).

(iii) Define a graph G'_m of order $p(\geq 6)$ as follows: for p, m ($1 \leq m \leq p - 2$), set

$$G'_m = K_1 + (P_{p-m-1} \cup \overline{K}_{m-1}),$$

where $\overline{K}_0$ stands for a null graph. Then we obtain G_m from G'_m by adding a new vertex x and joining x to one of the endvertices of P_{p-m-1} (see Figure 5). Then the graph G_m belongs to $\mathcal{E}_p(m,1)$. □

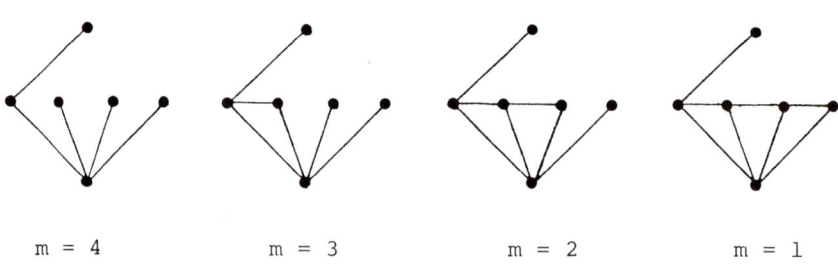

Figure 5. $p = 6$.

Before presenting Theorem 2, we introduce a definition and a lemma which are required in the proofs later.

Let G be a connected graph. An edge uv of G is said to be a *dominating edge* of G if any vertex w other than u or v in G is adjacent to either u or v.

Lemma 3. Let G be a connected graph with dominating edge $uv \in E(G)$ and with cutvertex x, then x is either u or v.

Proof. First we observe that for any $w \in V(G) - \{u, v\}$, uv is also a dominating edge of $G - w$. Then it follows immediately from the fact that a graph with a dominating edge is connected that every cutvertex of G is either u or v. □

Theorem 2. For $mn \geq 1$, $m \geq 2$, $C(m, n) \subset E$.

Proof. Let $G \in C(m, n)$, $mn \geq 1$, $m \geq 2$. Assume that $c(G) \geq 2$, $c(\bar{G}) \geq 1$ and that $v \in C(G)$, then

$$G - v = G_1 \cup G_2 \cup \ldots \cup G_n,$$

where k is the number of components of $G - v$.

Case 1. $k \geq 3$ for some $v \in C(G)$ or $k = 2$ and $|G_1|, |G_2| \geq 2$ for some $v \in C(G)$. If $k \geq 3$, $\bar{G} - v$ contains a spanning complete k-partite graph $K(|G_1|, \ldots, |G_k|)$, and thus $\bar{G} - v$ is a block. If $k = 2$ and $|G_1|, |G_2| \geq 2$, $\bar{G} - v$ contains a spanning complete bigraph $K(|G_1|, |G_2|)$, and so $\bar{G} - v$ is also a block. That is, $c(\bar{G} - v) = 0$. It follows from Lemma 1 (ii) that $v \in End(\bar{G})$ since $c(\bar{G}) = 1$ by the hypothesis. Therefore v is adjacent to exactly one vertex, say u, in $\bar{G}$, that is $uv \in E(\bar{G})$. However in this case, $G - u$ contains a spanning star with center v. Thus $c(G - u) \leq 1$. Since $c(G) \geq 2$ by the hypothesis, it follows from Lemma 1 (ii) that $u \in End(G)$. Thus, $G \in E$.

Case 2. $G - v = \{u\} \cup G_2$ for any $v \in C(G)$. Let v, v' be cut-vertices of G and u, u' be endvertices of G adjacent to v, v', respectively. Denote by H the graph obtained from G be removing four vertices u, u', v and v' (see Figure 6). Then $d_G(u, u') \geq 3$.

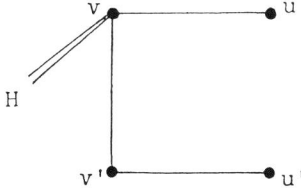

Figure 6.

This implies that the edge uu' is a dominating edge of $\bar{G}$. By the hypothesis, $\bar{G}$ has at least one cutvertex. Applying Lemma 3, we may assume that u is a cutvertex of $\bar{G}$. Since the closed neighbourhood $N[u']$ in $\bar{G}$ contains v and all the vertices of $V(H)$ in $\bar{G}$, v' must be an endvertex of $\bar{G}$. □

§2. A CLASSIFICATION OF BICONNECTED GRAPHS

We classify the set G of all biconnected graphs into 5 possible classes depending on the cutvertex-type $C(m,n)$, and also the endvertex-type $E(m,n)$ respectively, we give the structural characterizations for the following 6 classes of graphs:

$C(2,2)$, $C(2,1)$, $C(1,1)$ and E, $E(2,2)$, $E(m,1)$ $(m \geq 1)$.

Before presenting our results, we introduce a convenient notation in order to describe various kinds of families of graphs.

Let G, H be point-disjoint subgraphs of a graph F.

- $G === H$: The induced subgraph of $<V(G) \cup V(H)>$ in F has a complete bipartite subgraph with partite sets $V(G)$ and $V(H)$.

- $G ---- H$: There is a pair of adjacent vertices u, v with $u \in V(G)$, $v \in V(H)$ and at the same time there is a pair of nonadjacent vertices u, v with $u \in V(G)$, $v \in V(H)$ in F.

- $G \quad H$: No edge joins a vertex of G to a vertex of H.

- $G ---- H$: This notation stands for every possible adjacency relation between $V(G)$ and $V(H)$.

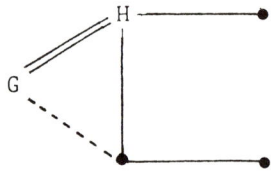

Figure 7.

We explain these notations by taking graphs illustrated in Figure 7 as an example. The family of graphs represented in Figure 7 are the 4 graphs in Figure 8 if $G \simeq K_3$, $H = K_1$.

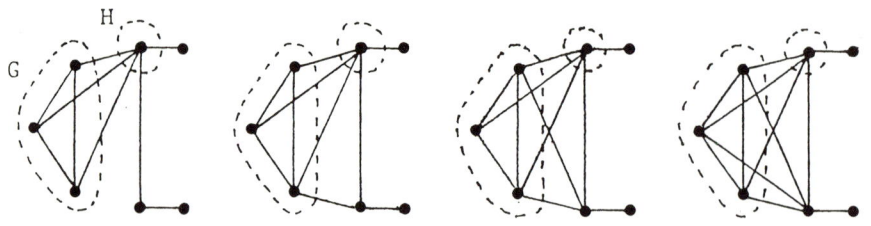

Figure 8.

Note that if G is the graph in Figure 9(a), then its complement $\bar{G}$ is one of the form in Figure 9(b), where F is an arbitrary graph and F' is its complement.

(a)　　　　　　　　　　　　(b)

Figure 9.

We are now ready to characterize graphs in E.

Theorem 3. Both G and $\bar{G}$ have at least one endvertex, i.e. $G \in \mathsf{E}$ if and only if G is one of the following graphs G_1 or G_2 in Figure 10, where F is an arbitrary graph (possibly a null graph).

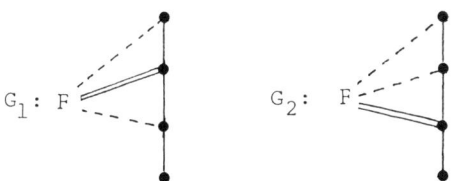

G_1: F　　　　　　　G_2: F

Figure 10.

Proof. First note that if G is one of the graphs of the form G_1 or G_2 in Figure 10, then $\bar{G}$ is the other. And so if G is represented in the form G_1 or G_2 then $G \in \mathsf{E}$. It thus suffices to show that if $G \in \mathsf{E}$ then G or $\bar{G}$ is represented as one of the graphs G_1 or G_2.

For any $G \in \mathsf{E}$, neither End(G) nor End($\bar{G}$) is empty. It follows from Lemma 2 that G contains an induced subgraph P_4 containing both u and v for any $u \in$ End(G) and any $v \in$ End($\bar{G}$). (We may assume without loss of generality that $uv \notin E(G)$.) Let F be the graph obtained from G by removing $V(P_4)$. Then

(1) $u \in \text{End}(G)$ if and only if every vertex of F is not adjacent to u in G.
(2) $v \in \text{End}(\bar{G})$ if and only if every vertex of F is adjacent to v in G.

Thus, G is a graph of the form G_1 or G_2 depending on $uv \notin E(G)$ or $uv \notin E(\bar{G})$ respectively. □

We now present two characterization theorems, the first one follows immediately from Theorems 2 and 3, the second from Theorem 3.

<u>Corollary 3.1</u>. A biconnected graph G is in $C(2,1)$ if and only if G or $\bar{G}$ is one of the graphs of the form G_1 or G_2 in Figure 11, where F is any graph.

Figure 11.

<u>Corollary 3.2</u>. For any given integer $m \geq 1$, a biconnected graph G is in $E(m,1)$ if and only if G or $\bar{G}$ is one of the following graphs G_1, G_2 or G_3 in Figure 12, where F is in G_1, G_2 and G_3 is any graph, and the induced subgraph $\langle N[F] \rangle$ in G_3 has no endvertex.

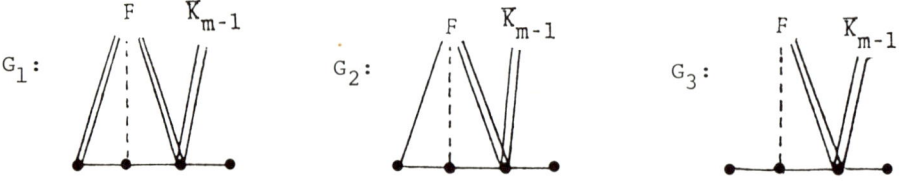

Figure 12.

Theorem 4. If G has at least 3 endvertices, then $\bar{G}$ has at most one endvertex, i.e.

$$E(m,n) = \emptyset \text{ for } m \geq 3, n \geq 2.$$

Proof. Let $G \in E$ be a graph with $e(G) \geq 3$. By Theorem 3, G is one of the graphs of the form G_1 or G_2 in Figure 10.

Case 1. G is a graph of the form G_1 in Figure 10. Since $e(G) \geq 3$, there exists a vertex $u \in V(F)$ such that $u \in \text{End}(G)$. This implies that $uv_1, uv_3 \notin E(G)$, i.e., $uv_1, uv_3 \in E(\bar{G})$, and so $v_1, v_3 u \notin \text{End}(\bar{G})$. Furthermore, since $uw, v_4 w \in E(\bar{G})$ for any $w \in V(F) - \{u\}$, we see that $w \notin \text{End}(\bar{G})$ for any $w \in V(F) - \{u\}$. Thus we obtain $\text{End}(\bar{G}) = \{v_2\}$, i.e. $e(\bar{G}) = 1$.

Case 2. G is a graph of the form G_2 in Figure 10. Since $e(G) \geq 3$, there exists a vertex $u \in V(F)$ such that $u \in \text{End}(G)$. Similar to Case 1, we obtain that $\text{End}(\bar{G}) = \{v_3\}$, and thus $e(\bar{G}) = 1$. By Case 1 and 2, we see that $e(\bar{G}) = 1$ if $e(G) \geq 3$. □

The following results are immediately from Theorems 2 and 4.

Corollary 4.1. For any integer m, n ($mn \geq 3$),
$$C(m,n) = \phi.$$
□

Corollary 4.2. The set $C(2,2)$ coincides with the set $E(2,2)$, furthermore any graph G in $C(2,2)$ is one of the graphs in Figure 13, where F is an arbitrary graph (possibly a null graph).

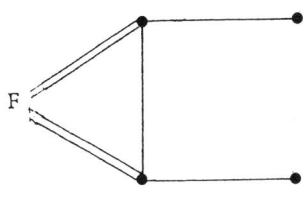

Figure 13.

Theorem 5. $C(1,1) \subset E \cup E(1,0)$.

Proof. We show that $G \in E(1,0)$ provided that $G \in C(1,1)$ and $G \notin E$. Assume that $e(G) = 0$ and $C(G) = \{v\}$, then

$G - v = G_1 \cup G_2 \cup \ldots \cup G_n$, where G_i ($i = 1, 2, \ldots, n$) is a component of $G - v$. Denote by p_i the order of G_i, then $p_i \geq 2$ for each i ($1 \leq i \leq n$), since $e(G) = 0$. Thus, $\bar{G} - v$ contains a spanning complete n-partite graph $K(p_1, p_2, \ldots, p_n)$. Thus $\bar{G} - v$ is a block and so $c(\bar{G} - v) = 0$ by the assumption. It follows from Lemma 1 (ii) that $v \in \text{End}(\bar{G})$, since $c(\bar{G}) = 1$ by the assumption. Therefore, we obtain

$$G \in E(1,0).$$ □

The structural characterization for graphs of $C(1,1)$ is derived from Theorem 5.

<u>Corollary 5.1</u>. A biconnected graph G in $C(1,1)$ if and only if G or $\bar{G}$ is one of the graphs in the form in Figure 14, where F is an arbitrary graph in Figure 14(a), and both F and H are graphs such that neither F nor $H - (N(u) \cup N(w))$ has an isolated vertex.

(a)　　　　　　　　　　　　(b)

Figure 14.

<u>Proof</u>. First we observe that the graphs of Figure 14(a) and Figure 14(b) are in $C(1,1)$.

In order to prove the necessity, we divide our proof into two cases according to Theorem 5.

<u>Case 1</u>. $G \in E \cap C(1,1)$. In this case, it follows immediately from Theorem 3 that G or $\bar{G}$ must be a graph of the form in Figure 14(a).

__Case 2.__ $G \in E(1,0) \cap C(1,1)$. Assume $\bar{G}$ has one endvertex v and G has no endvertex. Let u be the vertex adjacent to v in $\bar{G}$ and w be a vertex adjacent to u in G. Since $u \notin \text{End}(G)$, $u \notin C(G)$ by the assumption and Proposition 2, respectively, we obtain that $C(G - u) = C(G)$ by Lemma 1. Since v is a dominating vertex of $G - u$, $C(G - u) = C(G) = \{v\}$. Let H' be the connected component of $G - v$ containing the vertex u. Set

$$H = H' - u - w \quad \text{and} \quad F = G - H - v.$$

Then H is not a null graph for $\deg_G(u) = \deg_{H'}(u) \geq 2$ and also F is not a null graph for v is a cutvertex of G. Furthermore, since G has no endvertex, neither $H - N(u,w)$ nor F has an isolated vertex. Thus if $G \in E(1,0) \cap C(1,1)$ then G must be a graph of Figure 14(b). □

__Proposition 5.1.__ $C(0,0) \subset E(0,0)$.

__Proof.__ If G has an endvertex, then G has a cutvertex. □

From the results obtained above, we can classify a set G of biconnected graphs as in Table 2.

REFERENCES

[1] M. Behzad, G. Chartrand and L. Lesniak-Foster, *Graphs and Digraphs*, Prindle, Weber & Schmidt, (1979), Reading.

[2] F. Harary, *Graph Theory*, Addison-Wesley (1969), Reading.

[3] F. Harary, *Structural duality*, Behavioural Sci. 2 (1957) 225-265.

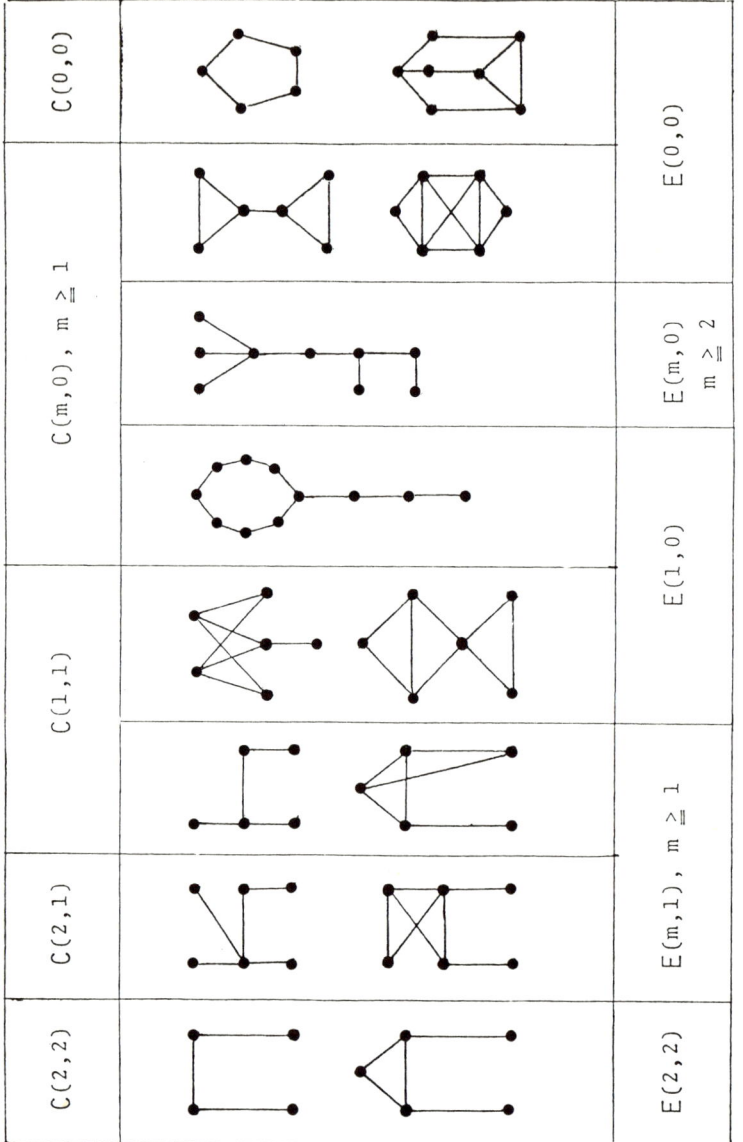

Table 2. A classification of a set G of biconnected graphs.

LINE GRAPHS AND THEIR CHROMATIC POLYNOMIALS

RUTH A. BARI

George Washington University
Washington, D.C. 20052
U.S.A.

Let G be a connected (p,q)-graph with $q > 0$, and let $L(G)$ be the line graph of G. Denote the chromatic polynomial of G by $P(G,\lambda)$, and the line chromial of G, that is, the chromatic polynomial of $L(G)$, by $P_L(G;\lambda)$. Since $L(G)$ has relatively many lines, it is most convenient to compute the line chromial of G in factorial form. A useful method for computing $P_L(G,\lambda)$ in this form is to multiply the partition matrix of the lines of G by the adjacency vector of $L(G)$. By this method, certain relations are derived between the points and lines of G and the coefficients of $P_L(G,\lambda)$. An appendix gives the coefficient vectors for the factorial forms of $P(G,\lambda)$ and $P_L(G,\lambda)$ for all connected (p,q)-graphs with $p \leq 5$ and $0 < q \leq 8$.

1. Introduction

The definitions and notation in this paper are based on [2]. In addition, we will need the following definitions. Let G be a (p,q)-graph whose lines have been labelled with integers $1,2,\ldots,q$, and let $L(G)$, the line graph of G, be a (p_L,q_L)-graph. The *adjacency vector* of $L(G)$ is the column vector
$A_L = [a_{12},a_{13},\ldots,a_{1q},a_{23},a_{24},\ldots,a_{2q},\ldots,a_{q-1,q}]^T$, with entries a_{ij} for all pairs i, j such that $i < j \leq q$, where $a_{ij} = 1$ if lines i and j are adjacent, and a_{ij} is 0 otherwise. If π is a partition of the lines of G into k parts, $1 \leq k \leq q$, the *partition vector* of π is the vector $P = [p_{12},p_{13},\ldots,p_{1q},p_{23},p_{24},\ldots,p_{2q},\ldots,p_{q-1,q}]$, with entries p_{ij} for all $i < j \leq q$, where $p_{ij} = 1$ if lines i and j are in the same cell of π, and $p_{ij} = 0$ if i and j are in distinct cells of π. A λ-*coloring* of $L(G)$ is a function from the set of lines of G into the set $I_\lambda = \{1,2,\ldots,\lambda\}$, whose elements are called *colors*. A λ-coloring is *proper* if no two

adjacent lines are assigned the same color. Associated with each λ-coloring of $L(G)$, there is a partition of the lines of G into color classes. A *color class* is a set of independent lines of G, or a matching. A *color partition* of $L(G)$ is a partition of the lines of G into color classes. If a color partition π has k color classes, the number of proper λ-colorings associated with π is $\lambda^{(k)} = \lambda(\lambda-1)(\lambda-2)\ldots(\lambda-k+1)$. Thus, if $P_L(G,\lambda)$ denotes the *line chromial of* G, i.e. the chromatic polynomial of $L(G)$, the set of all color partitions of the lines of G leads to the factorial form of $P_L(G,\lambda)$,

$$P_L(G,\lambda) = b_0 \lambda^{(q)} + b_1 \lambda^{(q-1)} + b_2 \lambda^{(q-2)} + \ldots + b_{q-1} \lambda^{(1)},$$

where b_i is the number of partitions of the lines of G into $q-i$ color classes.

The *Stirling numbers of the first kind*, denoted $s(n,r)$, are defined by the relation $\lambda^{(n)} = \lambda(\lambda-1)\ldots(\lambda-n+1) = \sum_{r=0}^{n} s(n,r)\lambda^r$. It follows from this definition that $s(n,n) = 1$, $s(n,n-1) = -\binom{n}{2}$, $s(n,1) = (-1)^{n-1}(n-1)!$, and $s(n,0) = 0$. The $q \times q$ *Stirling matrix of the first kind* is the matrix $S_{1,q}$ whose entries are Stirling numbers of the first kind, where

$$S_{1,q} = \begin{vmatrix} s(q,q) & s(q,q-1) & s(q,q-2) & \ldots & s(q,1) \\ 0 & s(q-1,q-1) & s(q-1,q-2) & \ldots & s(q-1,1) \\ 0 & 0 & s(q-2,q-2) & \ldots & s(q-2,1) \\ \cdot & \cdot & \cdot & \ldots & \cdot \\ 0 & 0 & 0 & \ldots & s(1,1) \end{vmatrix}$$

Let $P_L(G,\lambda) = \sum_{i=0}^{q-1} b_i \lambda^{(q-i)} = \sum_{i=0}^{q-1} c_i \lambda^{q-i}$ be the line chromial of G in factorial and standard forms respectively, and let $B = [b_0, b_1, \ldots, b_{q-1}]$ and $C = [c_0, c_1, \ldots, c_{q-1}]$ be their respective coefficient vectors. Then it is clear that $B \cdot S_{1,q} = C$.

2. **Counting the color partitions of $L(G)$.**

The following approach to counting color partitions is a modification of the method introduced by O'Connor [3].

Theorem 1. Let π be an arbitrary partition of the lines of G into k parts. Let P be the partition vector of π, and let A_L be the adjacency vector of $L(G)$. Then π is a color partition of $L(G)$ if and only if the scalar product $P \cdot A_L = 0$.

Proof. Suppose π is a color partition of $L(G)$. Then if lines i and j are in the same cell of π, they are independent, so $p_{ij} = 1$ implies $a_{ij} = 0$, and $a_{ij} = 1$ implies $p_{ij} = 0$ for all pairs i,j of lines of G. Thus $P \cdot A_L = 0$. Conversely, if π is not a color partition of L(G), then some cell of π contains an adjacent pair of lines, i and j. Thus $p_{ij} = 1$ and $a_{ij} = 1$, so $p_{ij} a_{ij} = 1$, and $P \cdot A_L \neq 0$. □

Suppose that A_L has exactly r 0-entries. Then G has r pairs of non-adjacent lines, or 2-line matchings.

Let $M = \{m_1, m_2, \ldots, m_r\}$ be the set of 2-line matchings of G. Then every matching of G is either a single line or a union of 2-line matchings.

A partition π is an *M-partition* of the lines of G if each cell of π contains either a single line or a union of 2-line matchings of G. The *partition matrix* P_M of L(G) is the matrix whose rows are the partition vectors of those M-partitions of L(G) which have at least Δ parts. In P_M, a partition vector is said to be at *level* k if it represents a k-part M-partition of L(G). The vectors at level $q, q-1, \ldots, \Delta$ appear in that order in P_M, but the order of the vectors within the same level is arbitrary. Different levels are separated by dashed lines.

Example 1

$$G = \quad \Delta = 3$$

$$A_L = \begin{array}{c} 12\ 13\ 14\ 15\ 23\ 24\ 25\ 34\ 35\ 45 \\ [1,\ 1,\ 0,\ 0,\ 1,\ 1,\ 1,\ 1,\ 0,\ 1] \end{array}$$

$$M = \Big[\{1,4\}, \{1,5\}, \{3,5\}\Big]$$

M-partitions

    5 parts: 1;2;3;4;5

    4 parts: 1,4;2;3;5   1,5;2;3;4   3,5;1;2;4

    3 parts: 1,4,5;2;3   1,4;3,5;2   1,3,5;2;4

Partition matrix $P_M(G)$

Level	Partition	12	13	14	15	23	24	25	34	35	45
5	1;2;3;4;5	0	0	0	0	0	0	0	0	0	0
4	1,4;2;3;5	0	0	1	0	0	0	0	0	0	0
4	1,5;2;3;4	0	0	0	1	0	0	0	0	0	0
4	3,5;1;2;4	0	0	0	0	0	0	0	0	1	0
3	1,4,5;2;3	0	0	1	1	0	0	0	0	0	1
3	1,4;3,5;2	0	0	1	0	0	0	0	0	1	0
3	1,3,5;2;4	0	1	0	1	0	0	0	0	1	0

$$P_M \cdot A_L = [\ 0\ |\ 0,0,0\ |\ 1,0,1]^T = D\ .$$

Theorem 2. The number of zeros in the vector $P_M \cdot A_L = D$ is the number of color partitions of $L(G)$. The number of zeroes at level $(q-k)$ of D is the coefficient b_k of $\lambda^{(q-k)}$ in the factorial form of the line chromial $P_L(G,\lambda)$.

Proof. By Theorem 1, each zero in D represents a color partition of $L(G)$. Since the partition vectors at level $q-k$ correspond to $(q-k)$-part partitions of the lines of G, the number of zeroes at level $q-k$ is the number of $(q-k)$-part color partitions of $L(G)$, that is, the coefficient b_k of $\lambda^{(q-k)}$ in $L(G)$. □

Theorem 3. Let G be a (p,q)-graph with points $v_1, v_2, \ldots, v_p$, and let $L(G)$ be the line graph of G. Then if G contains τ triangles, $L(G)$ contains $\tau_L = \tau - \sum_{i=0}^{p} \binom{d_i}{3}$ triangles, where d_i is the degree of v_i.

Proof. Corresponding to each triple of mutually adjacent lines of G there is a triple of mutually adjacent points of $L(G)$. Three lines of G are mutually adjacent only if they form a triangle in G or if they are incident with a common point of G. □

Theorem 4. Let G be a (p,q)-graph with points $v_1, v_2, \ldots, v_p$, and let $P_L(G,\lambda) = \sum_{i=0}^{n-1} b_i \lambda^{(n-i)}$. Let d_i denote the degree of v_i, Δ the maximum degree of G, and τ the number of triangles of G. Then

　　1. The coefficient b_i of $\lambda^{(n-i)}$ counts the number of $(n-i)$-part color partitions of the lines of G.

　　2. $n = q$, the number of lines of G, and $b_0 = 1$.

　　3. If $b_r \neq 0$, then $r \leq q - \Delta$.

　　4. $b_1 = \binom{q}{2} - \sum_{i=1}^{p} \binom{d_i}{2}$.

　　5. $b_2 = \binom{q_L}{2} - \sum_{i=1}^{p} \binom{d_i}{3} + b_1 \binom{q-1}{2} - s(q, q-2)$.

Proof. 1. The number of λ-colorings associated with each $(n-i)$-part color partition of $L(G)$ is $\lambda^{(n-i)}$. Hence b_i, the coefficient of $\lambda^{(n-i)}$, counts the number of $(n-i)$-part color partitions of $L(G)$.

　　2. The unique color partition of $L(G)$ with the greatest number of parts is that in which each line is a color class. Thus there are q color classes and $\lambda^{(q)}$ colorings of $L(G)$ associated with this partition, so $n = q$, and $a_0 = 1$.

　　3. The smallest number of color classes in a color partition of $L(G)$ is χ', the line chromatic number of G, so every

color partition has at least χ' color classes. Since, by Vizing's Theorem, $\chi' = \Delta$ or $\chi' = \Delta+1$, if $b_r \neq 0$, then $q-r \geq \Delta$ or $q-r \geq \Delta+1$. Hence $r \leq q - \Delta - 1 \leq q - \Delta$, so $r \leq q - \Delta$.

4. Let $B = [1, b_1, b_2, \ldots, b_r, 0, 0, \ldots, 0]$ be the coefficient vector of $P_L(G, \lambda)$, the line chromial of G expressed in factorial form, and $C = [1, c_1, c_2, \ldots, c_{q-1}]$ the coefficient vector of $P_L(G, \lambda)$ expressed in standard form. Since $B \cdot S_{1,q} = C$, we get $1 \cdot s(q, q-1) + b_1 \cdot s(q-1, q-1) = c_1$. But since $s(q, q-1) = -\binom{q}{2}$ and $s(q-1, q-1) = 1$, and it is known that $c_1 = -q_L = -\sum_{i=1}^{p}\binom{d_i}{2}$, we have $b_1 = \binom{q}{2} - \sum_{i=1}^{p}\binom{d_i}{2}$.

5. Again from the fact that $B \cdot S_{1,q} = C$, we get $1 \cdot s(q, q-2) + b_1 \cdot s(q-1, q-2) + b_2 \cdot s(q-2, q-2) = c_2$. But $s(q-1, q-2) = \binom{q-1}{2}$, $s(q-2, q-2) = 1$, and it is known that $c_2 = \binom{q_L}{2} - \tau_L = \binom{q_L}{2} - \left(\tau + \sum_{i=1}^{p}\binom{d_i}{3}\right)$.

Hence $s(q, q-2) - b_1 \binom{q-1}{2} + b_2 = \binom{q_L}{2} - \tau - \sum_{i=1}^{p}\binom{d_i}{3}$, so

$b_2 = \binom{q_L}{2} - \tau - \sum_{i=1}^{p}\binom{d_i}{3} + b_1\binom{q-1}{2} - s(q, q-2)$. □

<u>Example 2.</u> From Example 1, we know that if $G = $ [small graph figure]

$P_L(G, \lambda) = \lambda^{(5)} + 3\lambda^{(4)} + \lambda^{(3)}$. Then $q = 5$, $\Delta = 3$, $\tau_L = \tau + \sum_{i=1}^{p}\binom{d_i}{3}$

$= 1 + 2\binom{1}{3} + \binom{2}{3} + 2\binom{3}{3} = 1 + 0 + 0 + 2 = 3$. Therefore

$b_1 = \binom{5}{2} - \sum_{i=1}^{5}\binom{d_i}{2} = 10 - 2\binom{1}{2} + \binom{2}{2} + 2\binom{3}{2} = 3$

$b_2 = \binom{q_L}{2} - \tau - \sum_{i=1}^{p}\binom{d_i}{3} + b_1\binom{q-1}{2} - s(5,3)$

$= \binom{7}{2} - 1 - 2 + 3\binom{4}{2} - 35 = 1$.

References

[1] Bari, R.A., and Hall, D.W., Chromatic Polynomials and Whitney's Broken Circuits, J. Graph Theory 1 (1977), 269-275.

[2] Harary, F., Graph Theory, Addison Wesley, Reading, 1969.

[3] O'Connor, M.G., A Matrix Approach to Graph Coloring, Honors Thesis, Mt. Holyoke College (1980).

[4] Riordon, J., An Introduction to Combinatorial Analysis, Wiley, New York, 1967.

APPENDIX

Chromials and Line Chromials of Connected Graphs with at most Five Points and Eight Lines

In the following table, the coefficient vectors of the factorial form of $P(G,\lambda)$ and $P_L(G,\lambda)$ are given for each connected (p,q)-graph with $p \leq 5$ and $0 < q \leq 8$.

(p,q)	G	L(G)	$P(G,\lambda)$	$(P_L(G,\lambda)$
(2,1)			[1,0]	[1]
(3,2)			[1,1,0]	[1,0]
(3,3)			[1,0,0]	[1,0,0]
(4,3)			[1,3,1,0]	[1,1,0]
(4,3)			[1,3,1,0]	[1,0,0]
(4,4)			[1,2,0,0]	[1,1,0,0]
(4,4)			[1,2,1,0]	[1,2,1,0]
(4,5)			[1,1,0,0]	[1,2,1,0,0]
(4,6)			[1,0,0,0]	[1,3,3,1,0,0]
(5,4)			[1,6,7,1,0]	[1,0,0,0]
(5,4)			[1,6,7,1,0]	[1,3,1,0]
(5,4)			[1,6,7,1,0]	[1,2,0,0]
(5,5)			[1,5,4,0,0]	[1,3,1,0,0]
(5,5)			[1,5,5,4,0]	[1,4,3,0,0]
(5,5)			[1,5,4,0,0]	[1,4,2,0,0]
(5,5)			[1,5,4,0,0]	[1,2,0,0,0]

Line graphs and their chromatic polynomials

(p,q)	G	L(G)	P(G,λ)	$(P_L(G,\lambda)$
(5,5)			[1,5,1,0,0]	[1,5,1,0,0]
(5,6)			[1,4,2,0,0]	[1,5,4,0,0]
(5,6)			[1,4,2,0,0]	[1,4,3,0,0,0]
(5,6)			[1,4,3,0,0]	[1,6,8,1,0,0]
(5,6)			[1,4,4,1,0,0]	[1,6,9,2,0,0]
(5,6)			[1,4,2,0,0]	[1,5,5,1,0,0]
(5,7)			[1,3,1,0,0]	[1,7,12,1,0,0,0]
(5,7)			[1,3,1,0,0]	[1,6,9,2,0,0,0]
(5,7)			[1,3,0,0,0]	[1,6,9,4,0,0,0]
(5,7)			[1,3,2,0,0]	[1,8,17,8,0,0,0]
(5,8)			[1,2,0,0,0]	[1,9,23,17,2,0,0,0]
(5,8)			[1,2,1,0,0]	[1,10,29,25,2,0,0,0]

LARGE GRAPHS WITH GIVEN DEGREE AND DIAMETER III

J.C. BERMOND, C. DELORME and G. FARHI

Université de Paris Sud, Orsay, bât. 490
France

The following problem arises in the study of interconnection networks: find graphs of given maximum degree and diameter having the maximum number of vertices. In this article we give a construction which enables us to construct graphs of given maximum degree and diameter, having a great number of vertices from small ones; in particular we obtain a class of graphs of diameter 3, maximum degree Δ and having about $8\Delta^3/27$ vertices.

I. INTRODUCTION

We are interested in the (Δ,D) graph problem, one of the problems arising in telecommunications networks (or microprocessor networks). Let $G = (X,E)$ be an undirected graph with vertex set X and edge set E. The *distance* between two vertices x and y, denoted by $\delta(x,y)$, is the length of a shortest path between x and y. The *diameter* D of the graph G is defined as $D = \max_{(x,y) \in X^2} \delta(x,y)$. The degree $d(x)$ of a vertex x is the number of vertices adjacent to x and Δ denotes the maximum degree of G. (For a survey on diameters see [1].)

The (Δ,D) graph problem is that of finding the maximum number of vertices $n(\Delta,D)$ of a graph with given maximum degree Δ and diameter D. This problem arises quite naturally in the study of interconnection networks: the vertices represent the stations (or processors), the degree of a vertex is the number of links incident at this vertex and the diameter represents the maximum number of links to be used to transmit a message. The problem seems to have been first set in the literature by Elspas [7]. Different contributions have been made in the 70's and the known results were summarized in Storwick's article [10]. These results have been recently improved by Memmi and Raillard [8] and Quisquatter [9].

A simple bound on $n(\Delta,D)$ is given by Moore (see [3 or 4]):

$n(2,D) \leq 2D+1$ and for $\Delta > 2$, $n(\Delta,D) \leq \frac{\Delta(\Delta-1)^D-2}{\Delta-2}$. The graphs satisfying the equality are called *Moore graphs*. It has been proved by different authors (see Biggs [3 chap.23]) that Moore graphs can exist only if $\Delta = 2$ (the graphs being the (2D+1) cycles) or if $D = 2$ and $\Delta = 3, 7, 57$ (for $\Delta = 3$ and $\Delta = 7$ there exists a unique Moore graph respectively Petersen's graph on 10 vertices, and Hoffman and Singleton's graph on 50 vertices, for $\Delta = 57$ the answer is not known).

The aim of this article is to give a new construction, which gives graphs having a great number of vertices in particular for small diameters; as a corollary we obtain that $\liminf_{\Delta \to \infty} n(\Delta,3) \geq \frac{8}{27}\Delta^3$.

This paper is to be considered as a companion of [2,5,6] the construction is based on a new product of graphs which is studied in [2] and in [5,6] are given infinite families of graphs of given degree and diameter. In [5] it was proved that $n(\Delta,D) \geq (\frac{\Delta}{2})^D$, which gives only $\frac{\Delta^3}{8}$ in the case of diameter 3.

II DEFINITIONS

2.1 The * product

Let $G = (X,E)$ and $G' = (X',E')$ be two graphs. Take an arbitrary orientation of the edges of G and let U be the set of arcs. Finally, for each arc (x,y) of U, let $f_{(x,y)}$ be a one to one mapping from X' to X'.

We define the product $G*G'$ as follows:
The vertex set of $G*G'$ is the cartesian product $X \times X'$. A vertex (x,x') is joined to a vertex (y,y') in $G*G'$ if and only if

either $x = y$ and $\{x',y'\} \in E'$,

or $(x,y) \in U$ and $y' = f_{(x,y)}(x')$.

Remark: $G*G'$ can be viewed as formed by $|X|$ copies of G' where two copies generated by the vertices x and y are joined if (x,y) is an arc and in that case they are joined by a perfect matching depending on (x,y).

Examples: (a) Let $G = K_2$, (1,2) being the oriented arc and let $G' = C_5$. In Fig.2.1.a, we have represented K_2*C_5 where $f_{(1,2)}(x') = x'$ and in Fig.2.1.b, K_2*C_5 where $f_{(1,2)}(x') = 2x'$ (mod.5).

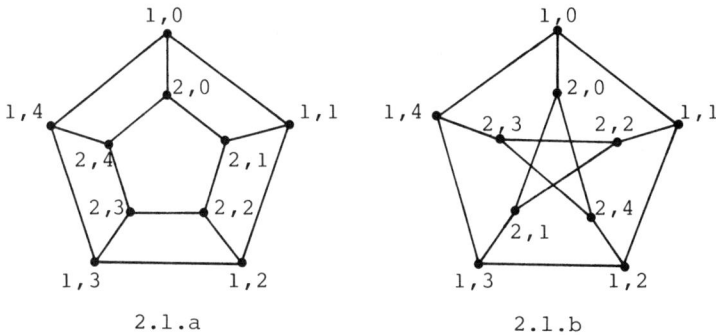

Fig. 2.1 : $K_2 * C_5$ with two different $f_{1,2}$

Note that the diameter of the graph of fig.2.1.a is 3, but the diameter of Petersen's graph (fig.2.1.b) is 2.

(b) If we choose for any arc (x,y) $f_{x,y}(x') = x'$ then G*G' is nothing else than the cartesian sum (called also cartesian product) of G and G' and in this case the diameter of G+G' is the sum of the diameters of G and G'.

Degrees and diameter of G*G'

If G is of maximum degree Δ and G' of maximum degree Δ', then G*G' has as maximum degree $\Delta + \Delta'$. If G is of diameter D and G' of diameter D', then the diameter of G*G' is less than or equal to D + D' and a clever choice of the functions $f_{(x,y)}$ can give a small diameter.

2.2 The property P*

A graph G = (X,E) is said to have the property P* if G has diameter at most 2 and if there exists an involution f of X (i.e. f^2 is the identity of X), such that for every vertex x of G we have $X = \{x\} \cup \{f(x)\} \cup \{f(\Gamma(x))\} \cup \{\Gamma(f(x))\}$.

Examples: In figure 2.2 are shown three graphs satisfying property P* with respectively 5,8 and 9 vertices. Other examples will be given in section 4.

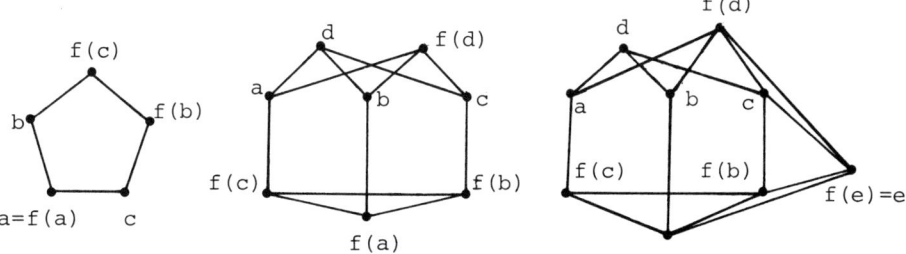

Figure 2.2

2.3 The property P

A graph G will be said to satisfy property P, if any pair of vertices at distance D (where D is the diameter of G) can be joined by a path of length $D+1$.

A trivial example is given by the complete graph K_n, $n \geq 3$; other examples will be given in section 4.

III THE MAIN THEOREM

<u>Theorem</u>: Let G be a graph of diameter $D \geq 1$, satisfying property P and let G' be a graph satisfying property P^* (with involution f). Then $G*G'$, with $f_{(x,y)}(x') = f(x')$ for every arc (x,y) of an arbitrary orientation of G, is a graph of diameter at most $D+1$.

<u>Proof</u>: Any two vertices (x,x') and (y,y') of $G*G'$ can be connected with a path of length at most $d_G(x,y) + 2$ (as G' is of diameter at most 2 by property P^*) and therefore if $d_G(x,y) < D$, then (x,x') and (y,y') are at distance at most $D+1$.

Now let x and y be at distance D and let z be the vertex preceding y on a path of length D between x and y. Let $z' = f^{D-1}(x')$. Then the vertex (z,z') is at distance $D-1$ of (x,x'); therefore the vertices in $(z,\Gamma(z'))$ and the vertex $(y,f(z'))$ are at distance at most D from (x,x') and the vertices in $(y,f(\Gamma(z')))$ and $(y,\Gamma(f(z')))$ at distance at most $D+1$ of (x,x'). Finally by using the path of length $D+1$ between x and y (by property P) the vertex $(y, f^{D+1}(x')) = (y,z')$ is at distance at most $D+1$ from (x,x') in $G*G'$. According to the fact that by property P^* the vertex set of G' is equal to $\{z'\} \cup \{f(z')\} \cup \{f(\Gamma(z'))\} \cup \{\Gamma(f(z'))\}$ every vertex of (y,G') is at distance at most $D+1$ from (x,x').

IV GRAPHS SATISFYING P* AND P

4.1 Graphs satisfying P*

We have seen three examples in 2.2 of degrees 2,3,4 and respectively 5,8,9 vertices. Note that a graph of degree Δ satisfying P^* has at most $2\Delta+2$ vertices. The next construction shows the existence of graphs of degree Δ on 2Δ vertices.

<u>Proposition</u>: For every integer Δ, there exists a graph G_Δ satisfying property P^*, of degree Δ and having 2Δ vertices.

<u>Construction</u>: Fig. 4.1.a shows such graphs with $\Delta = 1$ and $\Delta = 2$. Suppose we have constructed a graph G_Δ, satisfying P^*, of

degree Δ and such that the 2Δ vertices are A and f(A) with $|A| = |f(A)| = \Delta$ and such that $\Gamma(A) \supseteq f(A)$ and $\Gamma(f(A)) \supseteq A$ (that is satisfied by G_1 and G_2). Let $G_{\Delta+2}$ be the graph obtained from G_Δ by adding 4 vertices x,y,f(x),f(y) joined between themselves and to G as shown in Fig. 4.1.b. Then this graph (where the involution f extends that of G_Δ) has clearly degree $\Delta+2$, $2(\Delta+2)$ vertices and it is not difficult to show that it satisfies property P* and the hypothesis of induction.

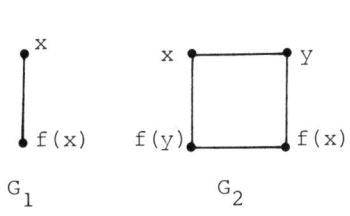
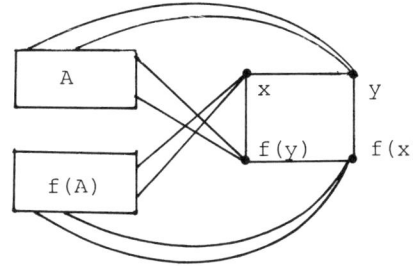

Fig. 4.1.a   Fig. 4.1.b

4.2 Graphs satisfying P

As already noticed the complete graphs K_n, $n \geq 3$, satisfy P. One can show also that Petersen's graph, Hoffmann-Singleton's graph, the products $K_n * X$, $n \geq 3$, where X is one of the graphs of Fig.2.2, satisfy property P.

An important class of graphs (which will be used in section 5) is formed by the graphs P_q of diameter 2 associated to projective planes. Let us recall, that if $q = p^r$ where p is a prime number, there exists a projective plane of order q with a polarity, that is a numbering of the points M_i, $1 \leq i \leq q^2 + q + 1$ and of the lines, D_j, $1 \leq j \leq q^2 + q + 1$ such that if $M_i \in D_j$ then $M_j \in D_i$. Thus $\underset{j \in D_i}{U} D_j$ contains all the points of the plane. Let P_q be the graph whose vertices are the points of the projective plane, the vertices M_i and M_j are joined if and only if $M_j \in D_i$. The properties of the numbering shows that P_q is of degree q+1; it has q^2+q+1 vertices and its diameter is 2. Now, let M_i and M_j be two vertices at distance 2, then $M_j \notin D_i$. As $|D_i \cap D_j| = 1$, there exists a vertex M_k, such that $M_k \in D_i$ and $M_k \notin D_j$; as M_k is at distance 1 from M_i and 2 from M_j, it is clear that M_i and M_j are joined by a path of length 3. Similarly one can prove that the quotient of Levi graphs of generalized quadrangles and hexagons, having a polarity, by that polarity have property P . (See [5].)

V APPLICATIONS

Let $G = K_n$ and $G' = X_8$ the graph on 8 vertices of degree 3 (see Fig. 2.2) then K_n*X_8 has 8n vertices, degree n+2 and diameter 2 . In particular with n = 3 (resp.4) we obtain graphs of diameter 2 , degree 5 (resp.6) on 24 (resp.32) vertices. The better graph of diameter 2 degree 6, known before had only 31 vertices (namely P_5) . By taking K_3*X_8 as graph G and again X_8 as G' we obtain a graph of diameter 3, degree 8, on 192 vertices (the better graph known before has 114 vertices). By taking for G the graph on 585 vertices of degree 9 and diameter 3 and for G' the graphs on 5(8,9) vertices of Fig.2.1 we give examples of graphs of diameter 4 of degree 11 (resp. 12,13) on 2925 (resp. 4680, 5265) vertices which are the best known now. Another application is the following result.

Theorem: $\lim\inf_{\Delta\to\infty} n(\Delta,3) \geq \frac{8}{27} \Delta^3$

Proof: Let G be the graph P_q defined in section 4, with q odd, q a prime power, of diameter 2, degree q+1 and on q^2+q+1 vertices which satisfies property P . Let $\Delta = \frac{3(q+1)}{2}$ and let $G' = G_{\Delta/3}$ be the graph of degree $\frac{\Delta}{3}$ on $\frac{2\Delta}{3}$ vertices satisfying property P* constructed in 4.1. By the main theorem $P_q*G_{\Delta/3}$ is a graph of diameter 3 degree $q+1+\frac{\Delta}{3} = \Delta$, having $(q^2+q+1) \frac{2\Delta}{3} = (4\frac{\Delta^2}{9} - \frac{2\Delta}{3} + 1)\frac{2\Delta}{3} = \frac{8\Delta^3}{27} - 4\frac{\Delta^2}{9} + 2\frac{\Delta}{3}$ vertices proving therefore the theorem.

VI GENERALIZATIONS

6.1 The property P_i

A graph G is said to have property P_i , if there exists a permutation f of its vertices, such that f^2 is an automorphism of G and that the graph G* , which has the same vertices as G and whose edges are the edges of G or their images by f , has diameter i .

6.2 Examples

The graph G of Fig.6, on 4 vertices, where f(x) = x+1 (modulo 4) has property P_1, G^* being the complete graph K_4.

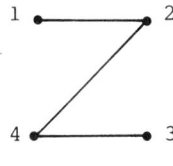

Fig. 6

The graph G, on 29 vertices, where the vertex x and y are joined if $y - x \equiv 1, -1, 7, -7 \pmod{29}$ and where $f(x) \equiv 12x \pmod{29}$ has property P_2.

If $4e+1$ is a prime power, consider the graph whose $4e+1$ vertices are the elements of the field GF_{4e+1}; two vertices x and y are adjacent if $x-y$ is a square in GF_{4e+1}. This graph has property P_1 (diameter 2 and degree 2e). (These graphs are known as Paley's graphs and are examples of strongly regular graphs.)

6.3 Theorem

Let G be a graph of diameter $D \geq i+1$ and let G' be a graph satisfying property P_i. Then $G*G'$, with $f_{(x,y)}(x') = f(x')$ for every arc (x,y) of an arbitrary orientation of G, has diameter $D+i$.

Proof: Let x and y be two vertices of G joined by a path P (not necessarily elementary) of length j. If this path contains a arcs of G and therefore $j - a$ arcs of $-G$, let $\varepsilon(P) = 2a-j$. We will prove that if there exists a path of length j' from $f^{\varepsilon(P)}(x')$ to y' in G'^*, with $j' \leq j$, then there exists a path of length $j+j'$ in $G*G'$ between (x,x') and (y,y'). The proof goes by induction on j (that is true for $j = 0$) and for j fixed by induction on $j'(j' \leq j)$; that is true for $j' = 0$ $(y, f^{\varepsilon(P)}(x'))$ being joined to (x,x') by the obvious path of length j. Suppose $j' > 0$ and let (z',y') be the last edge of the path between $f^{\varepsilon(P)}(x')$ and y' in G'^*. If $(z',y') \in G'$, we obtain a path of length $(j+j')$ from (x,x') to (y,y') by adding the arc $((y,z')(y,y'))$ to the path of length $j+j'-1$ from (x,x') to (y,z') (which exists by induction hypothesis). If $(z',y') \notin G'$ let z be the vertex preceding y on the path P of G; then $f^{\varepsilon(y,z)}(z')$ and $f^{\varepsilon(y,z)}(y')$ are joined by an edge in G' (recall that $\varepsilon(y,z)$ equals 1 or -1 according (y,z) belongs or not to G).

By induction hypothesis with $j-1$ and $j'-1$ ($j'-1 \leq j-1$) there exists in $G*G'$ a path of length $j+j'-2$ from (x,x') to $(z,f^{\varepsilon(y,z)}(z'))$ and then by adding the path of length 2 of $G*G'$ $(z,f^{\varepsilon(y,z)}(z'))$ $(z,f^{\varepsilon(y,z)}(y'))$ (y,y') we obtain a path of length $j+j'$ from (x,x') to (y,y').

Now let (x,x') and (y,y') be any two vertices of $G*G'$. There is a (not necessarily elementary) path P from x to y of length $j = D$ or $D-1$. Since $D \geq i+1$, $j \geq i$ holds. As G'^* is of diameter i, there exists a path of length $j' \leq i$ between $f^{\varepsilon(P)}(x')$ and y' in G'^*.

We have $j' \leq j$ (as $j' \leq i$ and $i \leq j$) and by the property proved above there exists a path of length $j+j' \leq D+i$ from (x,x') to (y,y').

6.4 Applications

By using the example of 6.2, we can build new good graphs. For example with the Paley's graphs we obtain infinitely many graphs of diameter 3 which have actually the highest known number of vertices, although the asymptotic value obtained in theorem of section 5 is not improved. By taking for G the graph of diameter 3, degree 9 on 585 vertices and for G' the Paley's graph on 13 vertices we obtain a graph of diameter 4, degree 15 on 7605 vertices. (This is in fact the best known value).

Remark: The condition on G, $D \geq i+1$ can be deleted if one adds extra conditions on G'; for example if G is of diameter 1 and G' satisfies P_1 and is of diameter 2, then $G*G'$ has diameter 2.

REFERENCES

[1] J.C. Bermond and B. Bollobás, Diameters in graphs: a survey, Proc. 12th Southeastern Conference on Combinatorics, Graph Theory and Computing, Baton Rouge 1981, to appear.

[2] J.C. Bermond, C. Delorme and G. Farhi, Large graphs with given degree and diameter II, to appear.

[3] N. Biggs, Algebraic graph theory, Cambridge University Press, Cambridge, England, 1974.

[4] B. Bollobás, Extremal graph theory, London Math. Soc. Monographs No. 11, Academic Press, London 1978.

[5] C. Delorme, Grands graphes de degré et diamètre donnés, to appear.

[6] C. Delorme and G. Farhi, Large graphs with given degree and diameter I, to appear.

[7] B. Elspas, Topological constraints on interconnection limited logic, Proc. 5th Symposium on switching circuit theory and logical design, I.E.E.E. 5-164 (1964), 133-197.

[8] G. Memmi and Y. Raillard, Some new results about the (d,k) graph problem, I.E.E.E. Trans. on Computers, to appear.

[9] J.J. Quisquatter: manuscript to be published.

[10] R.M. Storwick, Improved constructions techniques for (d,k) graphs, I.E.E.E. Trans. Computers, 19, (1970) 1214-1216.

DISTINGUISHING VERTICES OF RANDOM GRAPHS

BÉLA BOLLOBÁS

University of Cambridge
England

The *distance sequence* of a vertex x of a graph is $(d_i(x))_1^n$, where $d_i(x)$ is the number of vertices at distance i from x. The paper investigates under what condition it is true that almost every graph of a probability space is such that its vertices are uniquely determined by an initial segment of the distance sequence. In particular, it is shown that for $r \geq 3$ and $\varepsilon > 0$ almost every labelled r-regular graph is such that every vertex x is uniquely determined by $(d_i(x))_1^u$, where $u = \lfloor (\frac{1}{2}+\varepsilon)\frac{\log n}{\log(r-1)} \rfloor$. Furthermore, the paper contains an entirely combinatorial proof of a theorem of Wright [10] about the number of unlabelled graphs of a given size.

§0. INTRODUCTION

In this note we shall study two models of random graphs: $G(n,M)$ and $G(n,r\text{-reg})$. The first consists of all graphs of size M with vertex set $V = \{1,2,\ldots,n\}$ and the second is the set of all r-regular graphs with vertex set V. In both models all graphs are given the same probability. When dealing with the model $G(n,r\text{-reg})$ we shall assume that rn is *even*, so there are r-regular graphs of order n if $n \geq r+1$. For the basic facts concerning random graphs we refer the reader to [2, Ch.VII]. In particular, we say that *almost every* (a.e.) graph in a model has a certain property if the probability of having the property tends to 1 as the number of vertices tends to ∞.

Let G be $G(n,M)$ or $G(n,r\text{-reg})$. We are interested in the existence of indistinguishable vertices of a graph $G \in \mathcal{G}$ for two different definitions of being distinguishable. First we consider two vertices of G *distinguishable* if no automorphism of G maps one of the vertices into the other one. Thus $G \in \mathcal{G}$ consists of distinguishable vertices iff the automorphism group of G is trivial. To give the second definition of being distinguishable, given a vertex $x \in G$ write $d_i(x)$ for the number of vertices at distance i from x. Call $(d_i(x))_1^n$ the *distance sequence of* x and define two vertices to be *distinguishable* if they have different sequences. Clearly two vertices distinguishable in this sense are also distinguishable in the first sense.

A fundamental theorem of Wright [10] implies that if $M = M(n)$ is such that almost no graph G_M in $G(n,M)$ has two isolated vertices or two vertices of degree n-1, then almost every random graph G_M has trivial automorphism group. In fact the theorem states much more namely that the number of labelled graphs of order n and size M divided by the number of unlabelled graphs of order n and size M is asymptotic to n!. In §1 we give an entirely combinatorial proof of this result.

The rest of the paper is concerned with the problem of distinguishing vertices with the aid of their distance sequences. In §2 we treat the random graphs $G_M \in G(n,M)$ and in §3 we study random regular graphs of a fixed degree r. In both cases our aim is to use small portions of the distance sequences to identify the vertices.

§1. THE NUMBER OF UNLABELLED GRAPHS

Denote by $U_M = U_{n,M}$ the number of unlabelled graphs of order n and size $M = M(n)$ and write $L_M = L_{n,M}$ for the number of labelled graphs. We shall put $N = \binom{n}{2}$ so that $L_M = \binom{N}{M}$. Our aim is to show that under suitable conditions on M we have

$$U_M \sim L_M/n! = \binom{N}{M}/n! . \tag{1}$$

This is clearly considerably stronger than the assertion that the automorphism group of a.e. $G_M \in G(n,M)$ is trivial.

If the automorphism group of a graph of order n is trivial then the graph contains at most one isolated vertex and at most one vertex of degree n-1. It is easily seen (see e.g. Erdős and Rényi [6] or [7]) that this happens if and only if

$$\frac{2M}{n} - \log n \to \infty \quad \text{and} \quad \frac{2(N-M)}{n} - \log n \to \infty .$$

It is surprising that this simple necessary condition on M is sufficient to imply (1). This important result is due to Wright [10]; Earlier Pólya (see [8]) and Oberschelp [9] had proved (1) under considerably stronger conditions on M. Our aim is to give a combinatorial proof of this theorem.

Consider the symmetric group S_n acting on $V = \{1,2,\ldots,n\}$. For $\omega \in S_n$ let $G(\omega)$ be the set of graphs in G_M invariant under ω and put $I(\omega) = |G(\omega)|$. Then

$$\sum_{\omega \in S_n} I(\omega) = \sum_{G \in G_M} a(G) ,$$

where $a(G)$ is the order of the automorphism group of $G \in G_M$: $a(G) = |\text{Aut } G|$. Since G_M has exactly $n!/a(G)$ graphs isomorphic to a given graph $G \in G_M$,

$$U_M = \sum_{G \in G_M} (n!/a(G))^{-1} = \frac{1}{n!} \sum_{G \in G_M} a(G) = \frac{1}{n!} \sum_{\omega \in S_n} I(\omega).$$

For the identity permutation $1 \in S_n$ we have $I(1) = L_M$, so (1) holds iff

$$\sum_{\substack{\omega \in S_n \\ \omega \neq 1}} I(\omega) = o(L_M). \tag{2}$$

The number $I(\omega)$ depends only on the size of the orbits of ω acting on $V^{(2)}$, the set of N pairs of vertices, because a graph G_M belongs to $G(\omega)$ if and only if the edge set of G_M is the union of some entire orbits of ω acting on $V^{(2)}$.

Let A be a fixed set with a elements. We shall consider set systems $A = \{A_1, A_2, \ldots, A_s\}$ partitioning A: $A = \bigcup_{i=1}^{s} A_i$. Denote by $F(b;A) = F(b;A_1, \ldots, A_s)$ the collection of those b-element subsets of A which are the unions of some of the sets A_i. Thus $B \in F(b;A)$ iff $|B| = b$ and for every i we have $A_i \subset B$ or $B \cap A_i = \emptyset$. Suppose there are m_1 sets A_i of size 1, m_2 sets of size 2, ..., m_t sets of size t and no set with more than t elements. Note that these parameters satisfy

$$a = \sum_{j=1}^{t} j m_j. \tag{3}$$

The definition of $F(b;A)$ implies that if $f(b;A) = |F(b;A)|$ then

$$f(b;A) = f(b; m_1, m_2, \ldots, m_t) = \sum_{(i_j)} \binom{m_1}{b-k} \prod_{j=2}^{t} \binom{m_j}{i_j}, \tag{4}$$

where the summation is over all $(i_j)_2^t$ satisfying $0 \leq i_j \leq m_j$ and $k = k((i_j)_2^t) = \sum_{j=2}^{t} j i_j \leq b$, $b-k \leq m_1$.

If A' is obtained from A by decomposing an A_i into some sets (that is if the partition A' is a *refinement* of A) then clearly $F(b;A) \subset F(b;A')$, so

$$f(b;A) \leq f(b;A'). \tag{5}$$

In particular, if $m_1 + 2m_2 = a$ and $0 \leq k \leq m_2$ then

$$f(b; m_1, m_2) \leq f(b; m_1 + 2k, m_2 - k). \tag{6}$$

Furthermore, if A is the union of two disjoint systems, say A' and A'', then

$$f(b;A) = \sum_{i=0}^{b} f(b-i; A') f(i; A'').$$

Consequently if $m_j = m_j' + m_j''$ with $m_j' \geq 0$ and $m_j'' \geq 0$ then
$$f(b;m_1,\ldots,m_t) = \sum_{i=0}^{b} f(b-i;m_1',\ldots,m_t')f(i;m_1'',\ldots,m_t'') \ . \tag{7}$$
The last property of the function $f(b;m_1,\ldots,m_t)$ we need is not quite immediate so we state it as a lemma.

Lemma 1. Suppose $m_1 \leq a-2$. If $a-m_1$ is odd,
$$f(b;m_1,\ldots,m_t) \leq f(b;m_1+1,\tfrac{1}{2}(a-m_1-1))$$
and if $a-m_1$ is even then
$$f(b;m_1,m_2,\ldots,m_t) \leq f(b;m_1+2,\tfrac{1}{2}(a-m_1-2)) \ .$$

Proof. Since a set A_i with more than 3 elements can be partitioned into sets of size 2 and 3, by (5) we may suppose that A contains no sets of size greater than 3, that is $m_4 = m_5 = \ldots = m_t = 0$. Furthermore, if $m_3 = 0$ we are home since then $a-m_1 = 2m_2$ and by (6)
$$f(b;m_1,m_2,\ldots,m_t) = f(b;m_1,m_2) \leq f(b;m_1+2,m_2-1) \ .$$
Now suppose that $m_3 > 0$. We distinguish two cases according to the parity of m_3, which is also the parity of $a-m_1$. First suppose that m_3 is odd, say $m_3 = 2k+1 \geq 1$. We know from (7) that
$$f(b;m_1,m_2,2k+1) = \sum_{i=0}^{b} f(b-i;m_1,m_2)f(i;0,0,2k+1)$$
and
$$f(b;m_1+1,m_2+3k+1) = \sum_{i=0}^{b} f(b-i;m_1,m_2)f(i;1,3k+1) \ .$$
Hence it suffices to show that
$$f(i;0,0,2k+1) \leq f(i;1,3k+1) \ . \tag{8}$$
In order to show (8) we use (3) to write out the two sides explicitly. If the left-hand side of (8) is not zero then i is a multiple of 3. Furthermore, if $i = 6j$ then (8) becomes
$$\binom{2k+1}{2j} \leq \binom{3k+1}{3j}$$
and if $i = 6j+3$ then (8) is
$$\binom{2k+1}{2j+1} \leq \binom{3k+1}{3j+1} \ .$$
Both inequalities are obvious.

Secondly, if m_3 is even, say $m_3 = 2k+2$ then
$$f(b;m_1,m_2,2k+2) \leq f(b;m_1+1,m_2+1,2k+1) \leq f(b;m_1+2,m_2+3k+2) \ ,$$
completing the proof. □

Let us return to our central theme, relation (2). We shall prove it by estimating $I(\omega)$ in terms of the number of vertices fixed by ω. Denote by $S_n^{(m)}$ the set of permutations moving m vertices, that is fixing n−m vertices. Then $S_n^{(0)} = \{1\}$ and $S_n^{(1)} = \emptyset$. Furthermore, if $\omega \in S_n^{(m)}$ then the m vertices moved by ω can be selected in $\binom{n}{m}$ ways and there are at most m! permutations moving the same set of m vertices. Hence

$$|S_n^{(m)}| \le \binom{n}{m} m! = (n)_m. \tag{9}$$

Lemma 2. Let $2 \le m \le n$, $n \ge 3$, and denote by $N_1 = N_1(m)$ the integer satisfying

$$\binom{n-m}{2} + \frac{m+1}{2} \le N_1 \le \binom{n-m}{2} + \frac{m+4}{2}$$

for which $N_2 = (N-N_1)/2$ is also an integer. Then for $\omega \in S_n^{(m)}$ we have

$$I(\omega) \le f(M; N_1, N_2).$$

Proof. Suppose $\omega \in S_n^{(m)}$ as a permutation acting on $V^{(2)}$ has M_i orbits of size i, i = 1,2,...,n. Since a graph G_M belongs to $G(\omega)$ iff its edge set is the union of some entire orbits,

$$I(\omega) = |G(\omega)| = f(M; M_1, M_2, \ldots, M_n).$$

Hence if M_j' is chosen to be M_1+1 or M_1+2 so that $M_2' = (N-M_1')/2$ is an integer, then by Lemma 1 we have

$$I(\omega) \le f(M; M_1', M_2').$$

At most how large is M_1? If a permutation fixes two pairs of vertices, say $\{x,y\}$ and $\{y,z\}$, then it also fixes y, the common vertex, and so x and z as well. Consequently a pair fixed by $\omega \in S_n^{(m)}$ either consists of two vertices fixed by ω or it is disjoint from all other pairs fixed by ω. Hence

$$M_1 \le \binom{n-m}{2} + m/2$$

and so

$$M_1' \le N_1.$$

Therefore by inequality (6) we have

$$I(\omega) \le f(M; M_1', M_2') \le f(M; N_1, N_2). \qquad \square$$

It so happens that (2) can be proved fairly easily under a somewhat stronger assumption on M than necessary. Here we only state this result; for a proof we refer the reader to Wright [10]. As a consequence of this preliminary version of the main result in the proof of the main theorem we may assume that $M = O(n \log n)$.

Theorem 3. If

$$2n \log n \le M \le N - 2n \log n$$

then

$$U_M \sim L_M/n! . \qquad \square$$

Now we are ready to prove the fundamental theorem of Wright [10].

Theorem 4. Suppose $\psi(n) \to \infty$ and
$$\tfrac{1}{2}n(\log n + \psi(n)) \leq M \leq N - \tfrac{1}{2}n(\log n + \psi(n)) .$$
Then
$$U_M \sim L_M/n! .$$

Proof. The map sending a graph into its complement preserves the automorphism group and sets up a 1-1 correspondence between G_M and G_{N-M}. Hence in proving this result we may assume that $M \leq N/2$. Furthermore, because of Theorem 3 we may and shall assume that
$$\tfrac{1}{2}n \log n + \psi(n)n \leq M \leq 2n \log n .$$
We shall also assume that $n \geq n_o$ where n_o depends on the function $\psi(n)$ and is chosen so that all our inequalities hold. It will be easy to check that there is such an n_o.

Let $N_1 = N_1(m)$ and $N_2 = N_2(m) = \tfrac{1}{2}(N-N_1)$ be as in Lemma 2 and set
$$L(m,i) = \binom{N_1}{M-2i}\binom{N_2}{i} ,$$

$$L(m) = \sum L(m,i) ,$$
where the summation is over all i satisfying $0 \leq i \leq N_2$ and $M-N_1 \leq 2i \leq M$. If $\omega \in S_n^{(m)}$ then by Lemma 2 we have
$$I(\omega) \leq L(m) .$$
Hence by (2) and (9) the theorem follows if we show that
$$\sum_{m=2}^{n} (n)_m L(m)/L_M = o(1) . \tag{10}$$
For the sake of convenience let us recall the inequality N_1 satisfies:
$$m(n - \tfrac{m}{2} - 2) - 2 \leq N-N_1 \leq m(n - \tfrac{m}{2} - 2) - \tfrac{1}{2} . \tag{11}$$
Using standard estimates of binomial coefficients we find that
$$L(m,0)/L_M = \binom{N_1}{M}/\binom{N}{M} \leq (N_1/N)^M \leq \exp\{-2M \tfrac{m}{n}(1 - \tfrac{m+5}{2n})\} . \tag{12}$$
Set
$$\ell(m,i) = \frac{L(m,i)}{L(m,i-1)} = \frac{N-N_1-2i+2}{2i} \cdot \frac{(M-2i+2)(M-2i+1)}{(N_1-M-2i-1)(N_1-M+2i)}$$
and note that for a fixed value of m the ratio $\ell(m,i)$ is a decreasing function of i. In fact, if $1 \leq i \leq j$ then
$$\ell(m,j) \leq \tfrac{i}{j} \ell(m,i)$$
and so
$$L(m,j) \leq \frac{(\ell(m,1))^j}{j} L(m,0) . \tag{13}$$

In order to prove (10) we shall decompose the range of m into three intervals and show that the sum over each of these intervals is $o(1)$.

(a) Suppose that $2 \leq m \leq 10 \dfrac{n}{(\log n)^2}$. Then

$$\ell(m,1) = \dfrac{N-N_1}{2} \dfrac{M(M-1)}{(N_1-M+1)(N_1-M+2)}$$

$$\leq \dfrac{mn}{2} \dfrac{(2n \log n)^2}{(N-mn-M)^2} \leq 100 .$$

Since

$$\ell(m,i) \leq \dfrac{1}{i} \ell(m,1) ,$$

this shows that

$$L(m) \leq c\, L(m,0)$$

for some absolute constant c. Therefore by (12)

$$L(m)/L_M \leq c \exp\{-2M \dfrac{m}{n}(1 - \dfrac{m}{2n})\}$$

$$\leq c \exp\{-m(\log n + \psi(n))(1 - \dfrac{m}{2n})\}$$

$$\leq c \exp\{-m \log n - \psi(n)m - 2\dfrac{m^2}{n} \log n\}$$

$$\leq \exp\{-m \log n - \tfrac{1}{2}\psi(n)m\} .$$

Summing over the m in our range we find that

$$\sum_m (n)_m L(m)/L_M \leq \sum_m \exp\{-\tfrac{1}{2}\psi(n)m\} = o(1) .$$

(b) Now let us turn to the largest interval we shall consider:

$$10 \dfrac{n}{(\log n)^2} \leq m \leq n - n^{\frac{3}{4}} (\log n)^2 .$$

It is easily checked that in this range $\ell(m,1) \geq 4$.
Put

$$j = \lfloor 2\ell(m,1) \rfloor .$$

Then

$$\ell(m,j+1) \leq \dfrac{\ell(m,1)}{j+1} \leq \dfrac{\ell(m,1)}{2\ell(m,1)} = \tfrac{1}{2}$$

so, estimating rather crudely,

$$L(m) \leq 2 \sum_{i=1}^{j} \dfrac{(\ell(m,1))^i}{i} L(m,0) \leq 3(\ell(m,1))^{2\ell(m,1)} L(m,0) . \qquad (14)$$

Since

$$N_1/N \leq \left(\binom{n-m}{2} + m/2 + 2\right)/N \leq \left(\dfrac{n-m}{n}\right)^2 (1+n^{-\frac{1}{2}}) ,$$

we have
$$\log\{L(m,0)/L_M\} \le 2M \log \frac{n-m}{n} + M n^{-\frac{1}{2}}.$$
Furthermore,
$$N_1 - M > \frac{2}{5}(n-m)^2$$
so
$$\ell(m,1) \le \frac{n^2}{4} \frac{(2n \log n)^2}{(\frac{2}{5} n^{3/2}(\log n)^4)^2} < 7 n/(\log n)^6.$$
Hence by (14) we have
$$\sum (n)_m L(m)/L_M = o(1)$$
for the sum over the range at hand, if, say,
$$(m+2)\log n + 14 \frac{n}{(\log n)^6} \log n + 2M \log \frac{n-m}{n} + Mn^{-\frac{1}{2}} < 0. \qquad (15)$$
The derivative of the left-hand side with respect to m is
$$\log n - 2M/(n-m) < 0$$
so it suffices to check (15) for the minimal value of m, namely $m = \lceil 10n/(\log n)^2 \rceil$. Since
$$2M \ge n(\log n + \psi(n)) \quad \text{and} \quad \log \frac{n-m}{n} \le -\frac{m}{n},$$
the left-hand side of (15) is at most
$$2 \log n + 14n/(\log n)^5 - \psi(n)m + 2n^{\frac{1}{2}} \log n < 15n/(\log n)^5 - m < 0,$$
as required.

(c) Finally, suppose that
$$m \ge n - n^{\frac{3}{4}}(\log n)^2.$$
In this range the crudest estimates will do. Since $N_2 \le N/2$,
$$L(M,i)/L_M = \binom{N_2}{i}\binom{N_1}{M-2i}/\binom{N}{M} \le \frac{(N/2)^i}{i!} \frac{N_1^{M-2i}}{(M-2i)!} \frac{M!}{(N-M)^M} = h(m,i).$$
If $2(i+1) \le M$, we have
$$\frac{h(m,i+1)}{h(m,i)} = \frac{N}{2(i+1)} \frac{N_1^2}{(M-2i)(M-2i-1)} \ge \frac{n^4}{8 M^3} > 2,$$
as $N_1 \ge n/2$ holds for every m. Furthermore, at the maximal value of i, $\lfloor M/2 \rfloor$, we have
$$h(m,\lfloor M/2 \rfloor) \le n^4 M^{M/2}/(N-M)^{M/2} \le n^{-3n}.$$
Consequently in our range we have
$$\sum (n)_m L(m)/L_M \le \sum (n)_m 2h(m,\lfloor M/2 \rfloor) \le n^{-n}.$$

This completes the proof of our theorem. □

§2. DISTANCE SEQUENCES OF GRAPHS IN $G(n,M)$

Recall that the *distance sequence* of a vertex x is $(d_i(x))$, where $d_i(x) = \{z \in V : d(z,x) = i\}$. Our aim is to show that in a certain range of M almost every random graph G_M is such that its vertices are uniquely determined by their distance sequences. The distance sequence of an isolated vertex is $0, 0, \ldots$, so a graph with two isolated vertices does not have the property above. The next result shows that if M is not too large but it is large enough to ensure that almost no G_M contains two isolated vertices then almost every graph is such that its vertices are uniquely determined by the initial segments of their distance sequences.

Theorem 5. Suppose $\psi(n) \to \infty$ and
$$\tfrac{1}{2}n(\log n + \psi(n)) \leq M \leq n^{13/12}.$$
Then a.e. graph in $G(n,M)$ is such that for every pair of vertices (x,y) there is an $\ell \leq \ell_0 = 3\lceil (\log n/\log(M/n))^{\frac{1}{2}} \rceil$ with $d_\ell(x) \neq d_\ell(y)$.

Proof. For the sake of convenience we shall prove the analogous result for the model $G(n, P(\text{edge}) = p)$, where $p = 2M/n^2$. It will be clear that this does not change the essence of the proof only the calculations become a little more transparent.
Furthermore, the difficulties depend on the size of M: for M rather close to the lower bound we need a slightly different argument.

(a) Assume first that $pn/(\log n)^3 \to \infty$ and $p \leq 2n^{-12/13}$. Pick two vertices in V, say x and y. It suffices to show that in $G(n, P(\text{edge}) = p)$ the probability of $d_i(x) = d_i(y)$ for all $i \leq \ell_0 = 3\lceil (\log n/\log(M/n))^{\frac{1}{2}} \rceil$ is $o(n^{-2})$. Set

$$\Gamma_\ell(x,y) = \{z : d(z,x) = d(z,y) = \ell\},$$
$$\Delta_\ell(x) = \{z : d(z,x) = \ell, d(z,y) > \ell\},$$
$$\Delta_\ell(y) = \{z : d(z,y) = \ell, d(z,x) > \ell\},$$
$$\delta_\ell(x) = |\Delta_\ell(x)| \text{ and } \delta_\ell(y) = |\Delta_\ell(y)|.$$

Easy calculations show that in our range
$$(2pn)^{\ell_0} \leq n.$$

By Lemmas 3 and 5 of [4] we find that for some function $\varepsilon = \varepsilon(n) \to 0$ with probability $1 - o(n^{-3})$ we have for all $\ell \leq \ell_0$:

$$|\delta_\ell(x) - (pn)^\ell| \leq \varepsilon(pn)^\ell \text{ and } d_\ell(x) - \delta_\ell(x) \leq \varepsilon(pn)^\ell. \tag{16}$$

Similar relations hold for $\delta_\ell(y)$ and $d_\ell(y)$.

Now let $0 \le \ell \le \ell_0 - 1$ and suppose we are given the edges of a random graph $G_p \in G(n, P(\text{edge}) = p)$ spanned by the union of the set of vertices at distance at most $\ell+1$ from x and the set of vertices at distance at most ℓ from y. Denote by S the set of vertices not belonging to the union above and set $s = |S|$. [If (16) holds then clearly $s \ge n/(pn)$.]

What is the probability that $d_{\ell+1}(x) = d_{\ell+1}(y)$, conditional on the set of edges given so far? As the event $d_{\ell+1}(x) = d_{\ell+1}(y)$ conditional on the edges so far *determines* the number of vertices in S joined to some vertex in $\Delta_\ell(y)$, this probability is at most
$$\max_k b(k; s, 1-(1-p)^{\delta_\ell(y)}),$$
where $b(k; s, \gamma)$ denotes the kth term of the binomial distribution:
$$b(k; s, \gamma) = \binom{s}{k} \gamma^k (1-\gamma)^{s-k}.$$
Consequently the probability of $d_{\ell+1}(x) = d_{\ell+1}(y)$ conditional on (16) is at most
$$(pn)^{-(\ell+1)/2}.$$
Therefore the probability of $d_i(x) = d_i(y)$ for $i = 1, 2, \ldots, \ell_0$, is at most
$$o(n^{-3}) + \prod_{\ell=1}^{\ell_0} (pn)^{-\ell/2} = o(n^{-2}),$$
since
$$\prod_{\ell=1}^{\ell_0} (pn)^{\ell/2} \ge pn(pn)^{\ell_0^2/4}.$$
This completes the proof of the assertion in our range.

(b) Now assume that
$$\log n + \psi(n) \le pn \le (\log n)^4$$
for some function $\psi(n) \to \infty$. It is easily seen that under these assumptions a.e. random graph $G_p \in G(n; P(\text{edge}) = p)$ has the following properties:

(i) G_p has no isolated vertex,
(ii) every vertex has degree at most $3pn$,
(iii) for any two adjacent vertices there are at least ½ log n/log log n vertices adjacent to at least one of them,
(iv) for every path of length two there are at least ½ log n vertices adjacent to at least one of them,
(v) for every integer $\ell \le 20$, every set of ℓ vertices spans a subgraph of size at most ℓ.

Denote by A the event that a random graph G_p has the properties above. If A holds then we can proceed more or less as in part (a). Conditional on A, with probability at least $1 - n^{-3}$ we have

$$\sum_{1}^{\ell_o} (d_\ell(x) + d_\ell(y)) = o(n)$$

and

$$\min\{\delta_\ell(x), \delta_\ell(y)\} \geq (pn)^{\ell-3}$$

for all $\ell \leq \ell_o$. As in (a), these imply that, conditional on A, the probability of $d_i(x) = d_i(y)$ for all $i \leq \ell_o$ is at most

$$\prod_{i=1}^{\ell_o - 3} (pn)^{i/2} = o(pn)^{5\ell_o^2/21} = o(n^{-2}).$$

Hence the probability that for *some* pair of vertices (x,y) we have $d_i(x) = d_i(y)$ for all $i \leq \ell_o$ is at most

$$1 - P(A) + n^2 o(n^{-2}) = o(1). \qquad \square$$

Remarks

1. Theorem 5 is essentially best possible. One can show that a.e. random graph $G_M \in G(n,M)$ contains two vertices x and y such that $d_i(x) = d_i(y)$ whenever $i \leq (\log n/\log(M/n))^{\frac{1}{2}}$. Furthermore, it is easily seen that if $M = \lfloor n^{11/8} \rfloor$, say, then a.e. G_M has diameter 3 and a.e. G_M contains vertices x and y with $d_i(x) = d_i(y)$ for all i. [In fact $d_i(x) = d_i(y) = 0$ for $i \geq 4$.] Thus Theorem 5 does not hold for large values of M, though the bound $n^{13/12}$ could easily be improved.

2. The proof shows that the vertices can be distinguished by other parts of the distance sequence as well. For example if M is in our range, ℓ_1 and ℓ_2 are natural numbers,

$$(4M/n)^{\ell_1 + \ell_2} < n$$

and

$$(M/n)^{\ell_1 \ell_2} > n^5$$

then almost no random graph G_M contains two vertices x and y with $d_i(x) = d_i(y)$ for all $i : \ell_1 \leq i \leq \ell_2$.

§3. DISTANCE SEQUENCES OF RANDOM REGULAR GRAPHS

The study of random regular graphs was started only recently. The main reason for this is that the asymptotic number of labelled r-regular graphs of order n was found only in 1978 by Bender and

Canfield [1]. Even more recently in [3] a simpler proof was
given for the same formula; furthermore in [3] a model was given
for the set of regular graphs with a fixed vertex set, which can be
used to investigate random labelled regular graphs. In particular,
Bollobás and de la Vega [5] studied the diameter of random regular
graphs. Our approach here is fairly close to that of [3].
We shall give a brief description of the model mentioned above, in a
form suitable for the study of distance sequences. Let $r \geq 3$ be
fixed and denote by $G(n,r\text{-reg})$ the probability space of all
r-regular graphs with a fixed set of n labelled vertices. We
shall assume that rn is even and any two graphs have the same
probability. As customary, we shall say that *almost every* (a.e.)
r-regular graph has a certain property if the probability of having
the property tends to 1 as $n \to \infty$. Let $W_1, W_2, \ldots, W_n$ be disjoint r-element sets. A *configuration is a partition of* $W = \cup_1^n W_i$
into unordered pairs, called *edges*. If (a,b) is an edge of
a configuration F then we say that a is *joined to* b. Denote
by Ω the set of configurations and turn it into a probability
space by giving all configurations the same probability. Given a
configuration $F \in \Omega$, define $\phi(F)$ to be the multigraph having
vertex $V = \{W_1, W_2, \ldots, W_n\}$ in which W_i is joined to W_j if some
element of W_i is joined to some element of W_j in F. Clearly
every r-regular graph with vertex set V is of the form $\phi(F)$ for
some $F \in \Omega$ and $|\phi^{-1}(G)| = (r!)^n$ for every r-regular graph with
vertex set V. In fact, as shown in [3], for about
$e^{-(r^2-1)/4}$ of all configurations F is $\phi(F)$ an r-regular graph.
As an immediate consequence of this we see that a.e. r-regular graph
has a certain property if and only if a.e. configuration has the
corresponding property.

The property we are interested in is that of having two vertices
with the same distance sequence. Let F be a configuration. In
order to define the distance sequences of F we define the
distance $d(W_i, W_j) = d_F(W_i, W_j)$ between two classes W_i, W_j as the
minimal k for which there are classes $W_{i_0} = W_i, W_{i_1}, \ldots, W_{i_k} = W_j$
such that for every ℓ, $0 \leq \ell < k$, an edge of F joins an element
of W_{i_ℓ} to an element of $W_{i_{\ell+1}}$. Equivalently, $d_F(W_i, W_j)$ is
the distance between W_i and W_j in the graph $\phi(F)$. For $W_i \in V$
set $d_\ell(W_i) = \{W_j : d(W_i, W_j) = \ell\}$ and call $(d_\ell(W_i))_{\ell=1}^n$ the
distance sequence of W_i. Thus W_i has the same distance sequence

in a configuration F as in the graph $\phi(F)$.

Theorem 6. Let $r \geq 3$ and $\varepsilon > 0$ be fixed. Set $\ell_o = \lfloor (\tfrac{1}{2}+\varepsilon)\frac{\log n}{\log(r-1)} \rfloor$. Then a.e. r-regular labelled graph of order n is such that every vertex x is uniquely determined by $(d_i(x))_1^{\ell_o}$.

Proof. Analogously to the proof of Theorem 5, our aim is to show that the probability of two fixed classes W_i and W_j of a random configuration satisfying

$$d_\ell(W_i) = d_\ell(W_j), \quad 1 \leq \ell \leq \ell_o,$$

is $o(n^{-2})$. Let $1 \leq i < j \leq n$ be fixed integers. We shall select the edges of a random configuration one by one, taking those nearest to W_i and W_j first. This approach is closely modelled on that in Bollobás and de la Vega [5].

Suppose we have selected all edges at least one endvertex of which belongs belongs to a class at distance less than k from the set $\{W_i, W_j\}$. Take all classes at distance k from $\{W_i, W_j\}$ and one by one select pairs for those elements in these classes that have not been paired so far. Having done this consecutively for $k = 0, 1, \ldots, n-2$ and $n-1$, we have constructed the union of the components of W_i and W_j in a random configuration. In fact it is rather trivial that a.e. configuration is connected so after this sequence of operations we almost surely end up with the entire random configuration.

As in [5], we call an edge *indispensable* if it is the first edge that ensures that another class W_ℓ is at a certain distance from $\{W_i, W_j\}$. Equivalently, an edge is dispensable if each class containing an endvertex of the edge is either one of W_i and W_j or else it contains a vertex incident with an edge we have already chosen. Note that an edge joining two vertices of the same class is dispensable. What is the probability that the kth edge we select is dispensable? As the k-1 edges selected so far are incident with vertices in at most k+1 classes, this probability is at most

$$\frac{(k+1)(r-1)}{(n-1)r}.$$

Therefore the probability that more than 2 of the first $k_o = \lfloor n^{1/6} \rfloor$ edges are dispensable is at most

$$\binom{k_o}{3} \left(\frac{k_o}{n-k_o}\right)^3 = o(n^{-2}), \tag{17}$$

the probability that more than $\ell_1 = \lfloor n^{1/8} \rfloor$ of the first $k_1 = \lfloor n^{6/13} \rfloor$ edges are dispensable is at most

$$\binom{k_1}{\ell_1+1}\left(\frac{k_2}{n-k_2}\right)^{\ell_1+1} = o(n^{-2}) \tag{18}$$

and the probability that more than $\ell_2 = \lfloor n^{5/13} \rfloor$ of the first $k_2 = \lfloor n^{2/3} \rfloor$ edges are dispensable is at most

$$\binom{k_2}{\ell_2+1}\left(\frac{k_2}{n-k_2}\right)^{\ell_2+1} = o(n^{-2}). \tag{19}$$

Let A be the event that at most 2 of the first k_o edges are dispensable, at most ℓ_1 of the first k_1 and at most ℓ_2 of the first k_2. By (17)-(19) we find that the probability of A is $1 - o(n^{-2})$. Therefore it suffices to show that the probability of $d_\ell(W_i) = d_\ell(W_j)$ for $1 \le \ell \le \ell_o$ conditional on A is $o(n^{-2})$.

In order to estimate this conditional probability we describe another, slightly different way of selecting the edges of a configuration. As before, suppose we have selected all edges incident with vertices in classes at distance less than k from $\{W_i, W_j\}$. Partition the elements that are unpaired at this stage into edges and put an edge into our random configuration under construction if it satisfies one of the following three conditions. (i) One of the vertices incident with the edge belongs to a class at distance k from W_i, (ii) both vertices incident with the edge belong to classes at distance k from W_j, (iii) if we add the edges satisfying (i) then one endvertex of our edge belongs to a class at distance k from W_j and the other belongs to a class at distance $k+1$ from W_i. We call this the $(2k)th$ *operation*.

Suppose that after the completion of the $(2k)$th operation there are t_k vertices in the classes at distance k from W_j that have not been paired so far and there are s_k classes consisting entirely of unpaired elements. The $(2k+1)st$ *operation* consists of pairing the t_k vertices above with t_k elements of the s_k classes and adding all these pairs to our random configuration.

Having completed the $(2k+1)$st operation, we have determined all edges incident with vertices in classes at distance at most k from $\{W_i, W_j\}$ so we can proceed to the $(2k+2)$nd operation. Once again, if we perform consecutively the 0th, 1st, 2nd, ..., $(2n)$th operations, then we construct all edges in the components containing W_i and W_j.

We may assume that the ε appearing in the statement of the theorem satisfies $0 < \varepsilon < 1/8$. Note that then
$$2 + r(r-1)^{\ell_0} \leq 2 + rn^{\frac{1}{2}+\varepsilon} < n^{2/3}.$$
Now let us assume that A holds. It is easily seen that then for $k \leq \ell_0$ we have
$$t_k \geq (r-1)^{k-3} \text{ and } s_k \geq n/2. \tag{20}$$
(In fact both bounds are rather crude.) Note that $d_{k+1}(W_i)$ is determined before we begin the $(2k+1)$st operation so if $d_{k+1}(W_i) = d_{k+1}(W_j)$ has to hold then the pairs of the t_k vertices must belong to a given number of the s_k classes. Hence the probability that $d_{k+1}(W_i) = d_{k+1}(W_j)$ conditional on t_k and s_k satisfying (20) is at most the maximum of the probability that a t_k element subset of the union of s_k classes of r elements each meets exactly ℓ classes, the maximum being taken over all values of ℓ, t_k and r_k satisfying (20). The following lemma gives an upper bound for this maximum.

Lemma 7. Let $R = \bigcup_{1}^{s} R_i$ be a partitioning of an rs-element set into r-element sets, where $r \geq 3$ is fixed. Suppose $t \leq cs^{5/8}$ for some constant c. Denote by X_T the number of R_i's intersected by a random t-subset T of R. Then
$$\max_{\ell} P(X_T = \ell) \leq c_0 s^{1/2}/t$$
for some constant c_0.

Proof. We may assume without loss of generality that $t \geq s^{\frac{1}{2}}$. The probability that T has at least 3 elements in some R_i is at most
$$s\binom{r}{s}\binom{rs-3}{t-3}/\binom{rs}{t} \leq \frac{1}{6}t^3/s^2 \leq \frac{c^4}{6} s^{\frac{1}{2}}/t.$$
Since
$$P(X_T = \ell) \leq P(\max_i |T \cap R_i| \geq 3) + P(X_T = \ell \text{ and } \max_i |T \cap R_i| \leq 2),$$
the lemma follows if we show that
$$P_\ell = P(X_T = \ell \text{ and } \max_i |T \cap R_i| \leq 2) \leq c_1 s^{\frac{1}{2}}/t$$
for some constant c_1.

Clearly $P_\ell = 0$ if $\ell < t/2$ and otherwise
$$P_\ell = \binom{s}{\ell}\binom{\ell}{t-\ell} r^{2\ell-t}\binom{r}{2}^{t-\ell}/\binom{rs}{t}.$$

For $1 \le u \le (t-1)/2$ set
$$Q_u = \frac{(t-2u)(t-2u-1)}{(s-t+u+1)(u+1)} \cdot \frac{r-1}{2r} .$$
It is easily seen that if u is an integer then we have
$$P_{t-u-1}/P_{t-u} = Q_u .$$
Define u_o, $1 < u_o < t/2$, by
$$Q_{u_o} = 1 .$$
Then
$$u_o \sim \frac{t^2(r-1)}{2rs} = o(t^{\frac{1}{2}})$$
and P_{t-u} increases with u for $u \le u_o-1$ and decreases with u for $u \ge u_o-1$. Furthermore, easy calculations show that for $u_o - u_o^{\frac{1}{2}} \le u \le u_o-1$
$$Q_u \le 1 + c_2 u_o^{-\frac{1}{2}}$$
and for $u_o - 1 \le u \le u_o + u^{\frac{1}{2}}$
$$Q_u \ge 1 - c_2 u_o^{-\frac{1}{2}}$$
where c_2 is some constant. Hence
$$\min\{P_{t-u} : u_o - u_o^{\frac{1}{2}} \le u \le u_o + u_o^{\frac{1}{2}}\} \ge c_3 \max_\ell P_\ell \qquad (21)$$
for some positive constant c_3. Finally, by the definition of P_ℓ we have
$$\sum_\ell P_\ell \le 1 ,$$
so (21) implies that
$$\max_\ell P_\ell \le \frac{1}{c_3 u_o^{\frac{1}{2}}} \le c_u s^{\frac{1}{2}}/t$$
for some positive constant c_u. □

The proof of Theorem 6 is easily completed with the aid of Lemma 7. The probability that $d_\ell(W_i) = d_\ell(W_j)$ for $\ell \le \ell_o$ is at most
$$1 - P(A) + \prod_h^{\ell_o} \{c_o n^{\frac{1}{2}}/(r-1)^{\ell-3}\} ,$$
where $h = \lfloor \frac{1}{2} \frac{\log n}{\log(r-1)} \rfloor + 3$. Since
$$(r-1)^{\ell_o} \ge n^{(1+\varepsilon)/2} ,$$
the sum above is $o(n^{-2})$, completing the proof of our theorem. □

Remark. A slight variant of the proof above gives the following extension of the theorem:

Let $r \geq 3$ and $\frac{1}{2} < \delta < 1$ be fixed and set $h_0 = \lfloor \delta \frac{\log n}{\log(r-1)} \rfloor$, $h_1 = h_0 + \lceil \frac{6}{2\delta - 1} \rceil + 3$. Then a.e. r-regular labelled graph of order n is such that every vertex x is uniquely determined by $\{d_i(x) : h_0 \leq i \leq h_1\}$.

REFERENCE

[1] E.A. Bender and E.R. Canfield, The asymptotic number of labelled graphs with given degree sequences, J. Combinatorial Theory (A) 24 (1978) 296-307.

[2] B. Bollobás, Graph Theory - An Introductory Course, GTM vol.63, Springer-Verlag, New York - Heidelberg - Berlin, 1979.

[3] B. Bollobás, A probabilistic proof of an asymptotic formula for the number of labelled regular graphs, Europ. J. Combinatorics 1 (1980) 311-316.

[4] B. Bollobás, The diameter of random graphs, Trans. Amer. Math. Soc. to appear.

[5] B. Bollobás and W.F. de la Vega, The diameter of random regular graphs, to appear.

[6] P. Erdös and A. Rényi, On the evolution of random graphs, Publ. Math. Inst. Hungar. Acad. Sci. 5 (1960) 17-61.

[7] P. Erdös and A. Rényi, On the strength of connectedness of a random graph, Acta Math. Acad. Sci. Hungar. 12 (1961) 261-267.

[8] G.W. Ford and G.E. Uhlenbeck, Combinatorial problems in the theory of graphs, Proc. Nat. Acad. Sci. U.S.A. 43 (1957) 163-167.

[9] W. Oberschelp, Kombinatorische Anzahlbestimmungen in Relationen, Math. Ann. 174 (1967) 53-78.

[10] E.M. Wright, Graphs on unlabelled vertices with a given number of edges, Acta Math. 126 (1971) 1-9.

THE TRAIL NUMBER OF A GRAPH

BÉLA BOLLOBÁS and FRANK HARARY

University of Cambridge, England
and
University of Michigan, Ann Arbor, U.S.A.

The trail number of tr(G) of a graph G is the maximum length of a trail in it. Consider the set of all connected graphs with n vertices and m edges. We first easily determine the exact maximum value of tr(G) among these graphs. Then we give the exact minimum provided $m \leq 4(n-1)/3$. Finally we prove an asymptotically correct estimate of the minimum of tr(G) for all values of n and m.

1. INTRODUCTION

We follow a mixture of the terminology and notation of [1] and [3]. However, let G(n,m) be a *connected graph with* $n \geq 3$ *vertices and* m *edges*. A *trail* of G is a walk in which no edge occurs more than once. A *circuit* is a closed walk. An *eulerian trail* is a closed trail containing all the edges; a *covering trail* is such an open trail. The *trail number* tr(G) is the maximum length of a trail in G. Of course tr(G) = m if and only if G is eulerian or G has a covering trail. Buckley [2] independently proposed the definition of tr(G) and noted that this is not a ramsey function, a fact which does not concern us here. Our object is to investigate the maximum and minimum values of tr(G) for G(n,m). We first note that the maximum can be determined without any effort.

2. THE MAXIMUM TRAIL NUMBER OF A GRAPH

This question is sufficiently trivial that one can determine the answer quickly and exactly. We will see that this is not so for the minimum trail number.

<u>Theorem 1.</u> The maximum trail number of a graph G(n,m) is m if n is odd or $m \leq \binom{n}{2} - \frac{n}{2} + 1$; otherwise it is $\binom{n}{2} - \frac{n}{2} + 1$.

<u>Proof.</u> If n is odd then K_n is eulerian. The first m edges of an eulerian trail of K_n form a graph G(n,m) with trail number m.

If n is even, then $K_n - (\frac{n}{2} - 1)K_2$ is eulerian. Hence for $m \le \binom{n}{2} - \frac{n}{2} + 1$, the first m edges of an eulerian trail again form a graph $G(n,m)$ with trail number m. On the other hand, the trail number of a graph G of order n is at most $\binom{n}{2} - \frac{n}{2} + 1$. For if H is the subgraph formed by the edges of a trail in G, then H has at most two vertices of degree $n-1$, so

$$e(H) \le \frac{1}{2}\{2(n-1) + (n-2)^2\} = \binom{n}{2} - \frac{n}{2} + 1 . \qquad \square$$

3. MINIMUM TRAIL NUMBER FOR SPARSE GRAPHS.

Before considering the minimum trail number for arbitrary $G(n,m)$ graphs, we first determine it exactly for the *sparse graphs* defined here as satisfying $m = n+k$ where k is a fixed integer and n is large. Let $\mu(k,n)$ be the minimum value of $\text{tr}(G)$ for $G(n, n+k)$.

Theorem 2. The minimum trail number for the sparse graphs are given exactly by the following six equations:

$$\begin{array}{lll}
(-1) & \mu(-1,n) = 2 & n \ge 3, \\
(0) & \mu(0,n) = \begin{cases} 3 & n = 3, \\ 4 & n = 4,5,6, \\ 5 & n \ge 7, \end{cases} \\
(1) & \mu(1,n) = \begin{cases} 5 & n = 4,5,6, \\ 6 & n \ge 7, \end{cases} \\
(2) & \mu(2,n) = \begin{cases} 5 & n = 4 \\ 6 & n \ge 5, \end{cases} \\
(3) & \mu(3,n) = \begin{cases} 7 & n = 5,6, \\ 8 & n \ge 7, \end{cases} \\
(4) & \mu(k,n) = 2 & k \ge 4 \text{ and } n \ge 3k+4.
\end{array}$$

Proof. We now justify the assertions (-1) - (4).

(-1). As a connected graph with $m = n-1$ is a tree, the value of $\mu(-1,n)$ is attained when G is the star $K_{1,n-1}$ and is 2.

(0). A connected graph with $m = n$ is unicyclic, so we have $n \ge 3$. When $n = 3$, the only unicyclic graph is the triangle K_3, so $\text{tr} = 3$. There are just two unicyclic graphs with $n = 4$, both having $\text{tr} = 4$. Among the five unicyclic graphs G with $n = 5$, all but one have $\text{tr}(G) = 5$, the exception being the graph of Figure 1a which has $\text{tr} = 4$, so that $\mu(0,5) = 4$. There are exactly 10 unicyclic graphs of order 6, just two of which (Figure 1b,c) attain the smallest

possible trail number, 4. For n ≥ 7, one sees at once that
μ(0,n) = 5 as illustrated in Figure 1d.

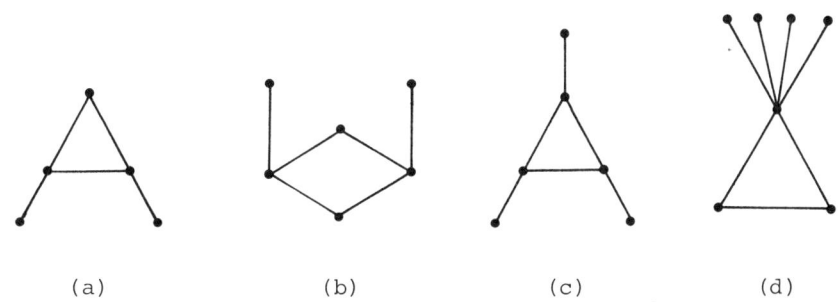

(a)   (b)   (c)   (d)

Figure 1. The extremal graphs for k = 0.

(1). If there are two edge-disjoint cycles in G, then there is
nothing to prove as we must have tr(G) ≥ 6. If not, then by
[1, p.120], G > T(K$_4$-e), that is G has a subcontraction to a
topological copy (subdivision) of K$_4$-e shown in Figure 2a. Since
K$_4$-e has a covering trail, G contains K$_4$-e as an induced subgraph, for otherwise tr(G) ≥ 6.

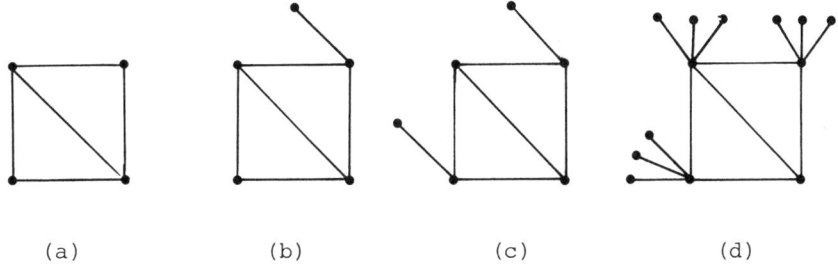

(a)   (b)   (c)   (d)

Figure 2. Extremal graphs for k = 1.

For n = 4, tr(K$_4$-e) = 5 as this graph has a covering trail. When
n = 5 and 6, the graphs of Figure 2b,c are the unique extremal
graphs which have trail number 5. For n ≥ 7, the extremal graphs
have the form of Figure 2d, with at least three endvertices; each
such graph has tr= 6.

(2). The complete graph K$_4$ is the only graph with n = 4 and
k = 2. Hence a fortiori it is the unique extremal graph for those
parameters and tr(K$_4$) = 5.

Now consider $n \geq 5$. First note that any longest circuit of G has length at most 5. Therefore there are no two edge-disjoint cycles. Hence, by Theorem 3.1 of [1, p.120], G must be homeomorphic to one of the three multigraphs $3K_2$ (three edges joining two vertices), K_4 or $K_{3,3}$.

Now $3K_2$ has too few edges and $K_{3,3} \supset C_6$, the hexagon, so we have $T(K_4)$, a topological K_4. But K_4 itself has $tr = 5$ so there is at most one edge of $T(K_4)$ which is subdivided and that edge is subdivided by at most one vertex of degree 2.

Let us recall that the *join* $G_1 + G_2$ of two disjoint graphs G_1 and G_2 is the union of G_1, G_2 and the complete bigraph $K(V_1, V_2)$; similarly $G_1 + G_2 + G_3 = (G_1+G_2) \cup (G_2+G_3)$, and so forth.

Now if just one edge of K_4 is subdivided by exactly one vertex, we obtain the graph $K_2 + \bar{K}_2 + K_1$ of Figure 3a which has $n = 5$ and $k = 2$. If this vertex of degree 2 is attached to a new endvertex, then we have $K_2 + \bar{K}_2 + K_1 + K_1$ (Figure 3b) with $n = 6$ and $k = 2$, both of these extremal graphs having $tr = 6$.

Otherwise for all $n \geq 5$, we have K_4 with a star $K_{1,r}$ attached to exactly one of its vertices, that is the graph $K_3 + K_1 + \bar{K}_r$ of Figure 3c. These are all the extremal graphs for $k = 2$.

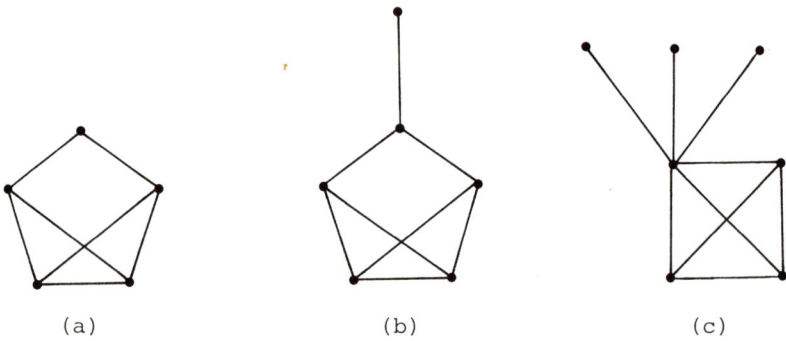

(a)   (b)   (c)

Figure 3. The extremal graphs for $k = 2$.

(3). As usual we first consider small n. The smallest possibility with $k = 3$ is $n = 5$ for which the unique extremal graph is the wheel $W_5 = .K_1 + C_4$ which has $tr = 7$.

When $n = 6$, there are just three extremal graphs: $K_1 + K_1 + C_4$, the graph obtained by adding a new endvertex to the center of W_5, and the two cubic graphs $K_{3,3}$ and $K_3 \times K_2$.

For $n \geq 7$, join $n-5$ independent vertices to a vertex of C_4 in the wheel $W_5 = K_1 + C_4$. The graph obtained has n vertices and $n + 3$ edges and its trail number is 8. Thus $\mu(n,3) \leq 8$.

We now prove that $\mu \geq 8$ whenever $n \geq 7$. Assuming that $\mu \leq 7$, any longest circuit has length at most 6.

Case 1. There are no two vertex-disjoint cycles.

We first note that there exist in this case two edge-disjoint cycles. Otherwise, by Corollary 3.3 of [1, p.120], G contains $T(K_{3,3})$. But as a longest circuit has length ≤ 6, the graph G must contain $K_{3,3}$ as a subgraph. Since $p \geq 7$, there exists a new vertex v adjacent to (a vertex of) $K_{3,3}$. Hence G has a trail of length 8.

Now we have two edge-disjoint cycles which are not vertex-disjoint. Both these cycles must be triangles because of the bound of 6 on the longest circuit length. Hence we have as a subgraph $K_3 \cdot K_3 = K_2 + K_1 + K_2$ and we denote its vertex set by W.

If P is a path of length 2 having no edge in $K_3 \cdot K_3$ but at least one vertex in W then $K_3 \cdot K_3 \cup P$ has two odd vertices so its trail number is 8. Consequently if we remove $E(K_3 \cdot K_3)$ from G, all remaining edges are independent. Therefore G[W], the subgraph induced by W, is the wheel W_5 and there exist at least two more (independent) edges, each having exactly one vertex in W. At least one of these edges is incident with a vertex of C_4 so once again our graph has trail number at least 8.

Case 2. There are two vertex-disjoint cycles.

As both these cycles must be triangles, and as they must be joined by just one edge (for otherwise G has an 8-trail, i.e., a trail of length 8), the subgraph implied by this case is $K_2 + K_1 + K_1 + K_2 = F$. Let c be one of the two cutvertices of this subgraph. If there is a v-c edge then we have an 8-trail, so there is none. Further, the other vertices, not in $W = V(F)$ must be independent for otherwise there is a 2-step path to a non-cutvertex of F resulting in an 8-trail. Also, no vertex $v \notin W$ can be adjacent to two vertices in W as this would yield either an 8-trail or two disjoint cycles one of which is C_4. Therefore all added edges must form stars at the non-cutvertices of F. As any longest circuit has length at most 6, the two additional edges to be added to the vertices of F to get $k = 3$ must give the triangular prism $K_3 \times K_2$ which has tr = 7. Finally, to have $n = 7$, we need one more vertex joined to any vertex of $K_3 \times K_2$, resulting in a graph with $tr = 8$, completing Case 2 and the proof of (3).

(4). We know from (3) that if $k \geq 4$ and $n \geq 7$ then $\mu(k,n) \geq 8$. Thus all we have to show is that for $k \geq 4$ and $n \geq 3k+4$ there is a graph $G = G(n, n+k)$ with trail number 8. Let G_o be a graph of order $n_o = 3(k+1) + 1$ obtained from $G_1 = (k+1)K_3$ by adding a new vertex x_o to G_1 and joining it by an edge to each triangle in G_1. Then G_o has $3(k+1) + k+1 = n_o + k$ edges and as x_o is a cutvertex, its trail number is 8. To obtain G add $n-n_o$ new vertices to G_o and join each to x_o. □

4. MINIMUM TRAIL NUMBER.

Our object is to derive close upper and lower bounds for the minimum trail number among the graphs $G(n,m)$. In the proof, we will require an elementary lemma. It is convenient to write $d(G) = d = 2m/n$ for the average of the degrees.

Lemma 3. Every graph $G(n,m)$ has a component of size at least $d(d+1)/2$.

Proof. Let G_1 to G_k where $G_i = G(n_i, m_i)$ be the components of $G = G(n,m)$. Define t_i by

$$m_i = t_i(t_i+1)/2,$$

so

and

$$t_i \leq n_i - 1$$

$$m_i \leq n_i t_i / 2.$$

Hence if for every i,

$$m_i < \frac{m}{n}\left(\frac{2m}{n} + 1\right) = d(d+1)/2,$$

then for every i,

$$m_i < \frac{m}{n} n_i$$

and so

$$m = \sum_1^k m_i < \sum_1^k \frac{m}{n} n_i = m,$$

a contradiction. □

Note that the bound $d(d+1)/2$ is best possible whenever both $d = 2m/n = t-1$ and $n/(d+1) = k$ are integers. For then kK_t has order n, size m, and each component has exactly $\binom{t}{2} = d(d+1)/2$ edges.

Lemma 4. If $G = G(n,m)$ satisfies $8d(d+1) \leq n(n-4)$, then it has a subgraph of size at least $d(d+1)$ which is the union of at most two components.

Proof. Let the components of G be $G_i = G(n_i, m_i)$, $i = 1,\ldots,k$ with $m_1 \geq m_2 \geq \ldots \geq m_k$. Then by Lemma 3,

$$m_1 \geq d(d+1)/2. \qquad (1)$$

Now if $m_1 \geq d(d+1)$, there is nothing to prove. Otherwise

$$m_1 < d(d+1) \qquad (2)$$

and so we apply Lemma 3 to $G - G_1$ to get

$$f(n_1, m_1) = m_1 + \frac{m-m_1}{n-n_1}\left(\frac{2(m-m_1)}{n-n_1} + 1\right) \leq m_1 + m_2. \qquad (3)$$

As $f(n_1, m_1)$ is at most the size of $G_1 \cup G_2$, it remains to show that it is at least $d(d+1)$, that is

$$g(n,m,n_1,m_1) = m_1 + 2\left(\frac{m-m_1}{n-n_1}\right)^2$$

$$+ \frac{m-m_1}{n-n_1} - \frac{2m}{n}\left(\frac{2m}{n} + 1\right) \geq 0. \qquad (4)$$

In proving (4) we shall not assume that the various parameters are integers, only that they satisfy the inequalities. Since $f(n_1, m_1)$ is a decreasing function of n_1, we may take $m_1 = n_1(n_1 - 1)/2$. Then (1) gives

$$n_1 \geq d + 1 = \frac{2m}{n} + 1, \qquad (5)$$

and by (2),

$$n_1(n_1 - 1) < 2d(d+1)$$

so the assumption on d implies $n > 4n_1$.

Now if m is as large as possible so that (5) is an equality, then $m_1 = \binom{d+1}{2}$ and $m_1 + m_2 \geq d(d+1)$ as required. Hence it suffices to show that

$$\frac{\partial}{\partial m} g(n,m,n_1,m_1) \leq 0 \qquad (6)$$

in the range allowed by (5). Now

$$n^2(n-n_1)^2 \frac{\partial g(n,m,n_1,m_1)}{\partial m} = 4(m-m_1)n^2 + (n-n_1)n^2 - 8m(n-n_1)^2 - 2n(n-n_1)$$

$$\leq 4\{(m-m_1)n^2 - 2m(n-n_1)^2\}$$

$$\leq 4\{-mn^2 + 4mnn_1\} = 4mn(4n_1-n) < 0,$$

so (6) holds. Then so does (4) and the lemma is proved. □

It is convenient to call a vertex *odd* or *even* in accordance with the parity of its degree. Then an *even graph* has all vertices even; in particular a connected even graph is eulerian. Let e(P) be the length of a path P. We have come to the final lemma needed in the proof of our main theorem.

<u>Lemma 5.</u> A graph G contains a forest F such that G-E(F) is an even graph.

<u>Proof.</u> Let E' ⊂ E(G) be a minimal set of edges whose deletion leaves an even graph. Then F = (V(G),E') is a forest. Indeed if C is a cycle in F then the parity of every vertex v ∈ V(G) is the same in G-E(F) as in G-E(F-E(C)) = G-(E'-E(C)), contradicting the minimality of E'.

<u>Theorem 6.</u> Let G = G(n,m) be a connected graph satisfying

$$8d'(d'+1) \leq n(n-4), \qquad (6)$$

where

$$d' = \frac{2m+2}{n} - 2.$$

Then

$$tr(G) \geq d'(d'+1) + 1.$$

<u>Proof.</u> By Lemma 5 our graph G contains an even subgraph G' with at least m' = m-n+1 edges. If G' is connected then it has a covering trail so

$$tr(G) \geq tr(G') \geq m-n+1 > d'(d'+1) + 1.$$

Now suppose that G' is disconnected. Since d' = 2m'/n satisfies (6), Lemma 4 implies that G' has a subgraph of size at least d'(d'+1) which is the union of two components, say H_1 and H_2. As G is connected it contains an H_1-H_2 path P whose internal vertices do not belong to $V(H_1) \cup V(H_2)$. Then $H_1 \cup H_2 \cup P$ is connected and has exactly two vertices of odd degree. Consequently

$$tr(G) \geq tr(H_1 \cup H_2 \cup P) = e(H_1 \cup H_2 \cup P) \geq d'(d'+1) + 1. \qquad \square$$

Corollary 7. If the average degree d of a connected graph $G = G(n,m)$ satisfies

$$d \leq \frac{n}{2\sqrt{2}} + 1,$$

then

$$tr(G) \geq (d-1)(d-2) + 1. \qquad \square$$

Remarks. (i) Theorem 6 is close to being best possible. For example if $n = kt+1$, $k \geq 2$ and $m = k\binom{t}{2}+k$ then there is a $G = G(n,m)$ such that

$$tr(G) = t^2 - t+2.$$

Such a graph G is obtained from k K_t by adding one vertex and k edges (see Figure 4). In particular, if $2k \geq t$ then the

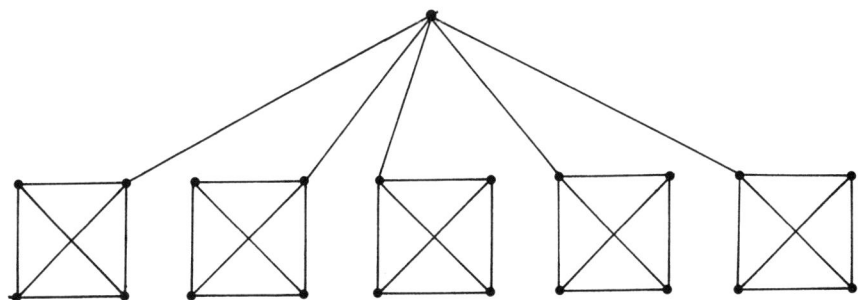

Figure 4. For $t = 4$ and $k = 5$ we obtain $G = G(21,35)$ with $tr(G) = 12$.

average degree d is at least $t-1$ and so

$$tr(G) \leq d(d+1) + 2.$$

Furthermore, if t is even then

$$tr(G) = t^2 - 2t + 4$$

so if $2k \geq t$ then

$$tr(G) \leq d^2 + 3.$$

Note that the graph defined in the proof of (4) of Theorem 2 is exactly this graph G for $t = 3$.

(ii) It is possible to strengthen Theorem 6 in various ways. For example one can prove that if $t \geq 3$ and is odd, and k is sufficiently large, then the minimum trail number of any $G(kt+1, k\binom{t}{2})$ is exactly $t^2 - t + 2$. However our proof is lengthy and far from elegant, so it is omitted.

REFERENCES

1. B. Bollobás, <u>Extremal Graph Theory</u>. Academic Press, London (1978).

2. F. Buckley, A ramsey property for graph invariants. <u>J Graph Theory</u> 4 (1980) 383-387.

3. F. Harary, <u>Graph Theory</u>. Addison-Wesley, Reading (1969).

MESSAGE GRAPHS

J.H. CONWAY and M.J.T. GUY

University of Cambridge
England

In this paper, the term v-*graph* will mean finite directed graph with constant invalence $v \geq 2$ and the same constant outvalence. We shall write $p \to q$ to mean that there is an edge directed from the vertex p to the vertex q. We shall think of such a graph as a system for transmitting messages, with $p \to q$ meaning that it is possible to send a message from p to q in a single step. Accordingly we shall call G a *cumulative d-step (message) graph* if for every pair of vertices p,q there is a directed path of at most d edges from p to q. In a similar way, we shall call G a *precisely d-step (message) graph* if for every p,q there is a path of *exactly* d edges directed from p to q. Finally, we shall call G *transitive* if its automorphism group is transitive on the vertices.

Our interest in message graphs was prompted by conversations with M. S. Paterson, who asked us how many vertices a transitive cumulative d-step 2-graph could have, and remarked that he knew of a transitive cumulative (3n-1)-step 2-graph with $n.3^n$ vertices. We found ourselves just as interested in the non-transitive cases.

We remark first that a precisely d-step v-graph can have at most v^d vertices, and a cumulative one at most $1+v+v^2+\ldots+v^d$. A graph which meets this bound will be called *tight*.

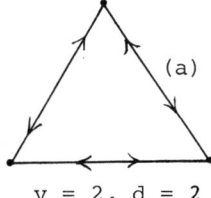

(a)

v = 2, d = 2

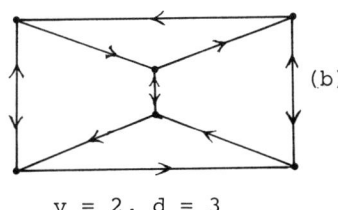

(b)

v = 2, d = 3

Figure 1.   Two precisely d-step v-graphs

Then we shall prove

Theorem 1. Tight precisely d-step graphs exist for all d.

Theorem 2. Tight cumulative d-step graphs exist only for d = 0 or 1.

Theorem 3. If there is a transitive precisely d-step v-graph with N vertices, then (for all n) there exists a transitive cumulative (dn+n-1)-step v-graph with $n \cdot N^n$ vertices.

From our point of view the graph mentioned by Paterson is that obtained by the construction of our Theorem 3 to the bi-directed triangle (Figure 1a). By applying it to the graph of Figure 1b instead, we obtain a cumulative (4n-1)-step graph with $n \cdot 6^n$ vertices, which is rather better, since $6^{1/4} > 3^{1/3}$.

Proof of Theorem 1. This is easy. We take the vertices of G as all length d sequences of letters from an alphabet of size v, with joining relation $a_0 a_1 \ldots a_{d-1} \to a_1 a_2 \ldots a_d$ for every choice of d+1 letters $a_0, a_1, \ldots, a_d$. (See Figure 2.)

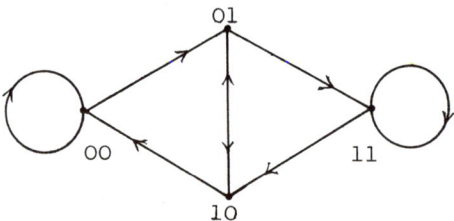

Figure 2. A tight precisely d-step v-graph for d = v = 2

Proof of Theorem 2. For d = 0, the graph is just a point joined to itself v times (Figure 3a), and for d = 1 a complete graph with each edge doubly-directed (Figure 3b). So the main problem (and indeed the main result of this paper) is to establish the non-existence for larger d.

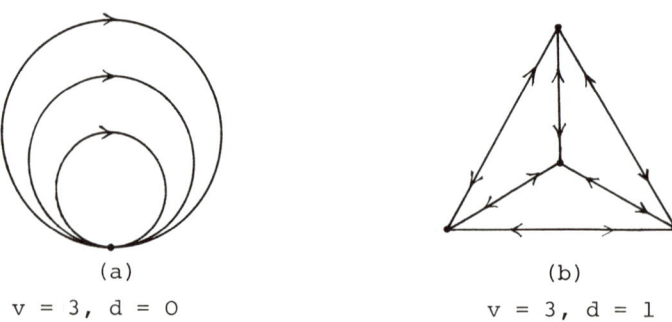

(a) v = 3, d = 0   (b) v = 3, d = 1

Figure 3. Tight cumulative d-step graphs

We consider the vector space of formal linear combinations of the vertices (and always refer it to the base consisting of the vertices). On this space we build the linear map α defined by $\alpha(p) = \sum q$ over all q for which $p \to q$. Then the tightness is expressed by the equation

$$1 + \alpha + \alpha^2 + \ldots + \alpha^d = J,$$

where J is represented by the all 1s matrix, so that $\alpha J = vJ$. Multiplying by $\alpha - v$, we deduce

$$(\alpha - v)(\alpha - \varepsilon)(\alpha - \varepsilon^2) \ldots (\alpha - \varepsilon^d) = 0,$$

where ε is a primitive $(d+1)$st root of unity, and since the numbers $v, \varepsilon, \varepsilon^2, \ldots, \varepsilon^d$ are distinct, that α is diagonalizable. It is easy to see that the multiplicity of the eigenvalue v is 1 (all coordinates of an eigenvector must equal the coordinate of maximal absolute value): let m_k be the multiplicity of ε^k ($1 \leq k \leq d$). Then by considering the trace of the maps $1, \alpha, \alpha^2, \ldots, \alpha^d$ we deduce

(0) $\quad 1 + m_1 + m_2 + \ldots + m_d = 1 + v + \ldots + v^d$,

(1) $\quad v + m_1 \varepsilon + m_2 \varepsilon^2 + \ldots + m_d \varepsilon^d = 0$, and so on up to

(d) $\quad v^d + m_1 \varepsilon^d + m_2 \varepsilon^{2d} + \ldots + m_d \varepsilon^{d^2} = 0$.

(All the matrices $\alpha, \alpha^2, \ldots, \alpha^d$ have zero diagonal.)

If we multiply the equation (i) by ε^{-ki}, and sum, we derive:

$$1 + \varepsilon^{-1}v + \varepsilon^{-2k}v^2 + \ldots + \varepsilon^{-dk}v^d + (d+1)m_k = 1 + v + \ldots + v^d,$$

which is equivalent to

$$\frac{v^{d+1} - 1}{\varepsilon^{-k}v - 1} + (d+1)m_k = \frac{v^{d+1} - 1}{v - 1}.$$

This, on the other hand, implies that ε^k is real for $k = 1, \ldots, d$, and so $d + 1 \leq 2$, which proves the Theorem. □

Proof of Theorem 3. We let G be the graph given in the hypotheses, and construct the new graph G_n as follows. The vertices of G_n are expressions of the form

$$(i | p_0 p_1 \ldots p_{n-1}),$$

where $i = 0, 1, \ldots, n-1$ and each p_k is a vertex of G. The joining relation is

$$(i | p_0 p_1 \ldots p_{n-1}) \to (i+1 | p_0 p_1 \ldots p_{i-1} q_i p_{i+1} \ldots p_{n-1})$$

where the subscripts are read modulo n, and $p_i \to q_i$. It helps to think of $p_0 p_1 \ldots p_{n-1}$ as a "tape", and of i as a "reading head" which at each step "advances" the symbol p_i currently scanned to one of the q_i for which $p_i \to q_i$, and then cycles one space. The graph is obviously transitive, since if π is an automorphism of G and t any integer, the map

$$(i|p_0 p_1 \ldots p_{n-1}) \to (i+t|\pi(p_0)\pi(p_1)\ldots\pi(p_{n-1}))$$

is an automorphism of G_n.

To see that it is a cumulative (dn+n-1)-step graph, let us suppose that we wish to find a path from $(i|p_0 p_1 \ldots p_{n-1})$ to $(j|q_0 q_1 \ldots q_{n-1})$. There is some t with $0 \leq t \leq n-1$ and $i+t \equiv j$ (modulo n), and so any t-step path from $(i|p_0 p_1 \ldots p_{n-1})$ will terminate at a vertex of the form $(j|r_0 r_1 \ldots r_{n-1})$. Since we have used $\leq n-1$ steps we still have dn steps available. We can therefore perform d complete cycles of our "reading head", in which it will "advance" each of $r_0, \ldots, r_{n-1}$ exactly d times, and so can be used to transform them into $q_0, q_1, \ldots, q_{n-1}$ respectively. This completes the proof. □

We end with several problems:

Problem 1. Is every tight precisely d-step v-graph isomorphic to the one we constructed in Theorem 1?

Problem 2. Is there a transitive tight d-step v-graph?

(Affirmative answers to both problems are impossible.)

Problem 3. We note that Figure 1b is also a cumulative 2-step 2-graph. Is there another cumulative d-step v-graph with $v+v^2+\ldots+v^d$ vertices?

Problem 4. Transitive precisely 4-step 2-graph with 12 or more vertices? (If so, application of Theorem 3 will produce a cumulative (5n-1)-step 2-graph with at least $n \cdot 12^n$ vertices, and we note that $12^{1/5} > 6^{1/4}$.)

Considerably more is known about similar problems concerning undirected graphs, see [1] and [2; Ch.IV].

REFERENCES

[1] J.-C. Bermond and B. Bollobás, Diameters of graphs - A survey, to appear.

[2] B. Bollobás, *Extremal Graph Theory*, London Math. Soc. Monographs, No.11, Academic Press, London, 1978.

SETS OF GRAPHS COLOURINGS

DAVID E. DAYKIN and PETER FRANKL

University of Reading, England
and
C.N.R.S., Paris, France

Let G be a finite graph and b, c be positive integers. We study the maximum number M of colourings of G so that adjacent *pairs* of G get the same colour at most b times. A colouring gives each edge, vertex of G one of c colours. Pair means:- *Case* e. Two edges. *Case* v. Two vertices. *Case* ev. Two edges or two vertices or an edge and a vertex. In all cases $\lim(b \to \infty)$ M/b is bounded.

1. INTRODUCTION

This paper has three cases e, v, ev. We have a fixed finite graph G and when we say *pair* we mean:- Case e. Two edges. Case v. Two vertices. Case ev. Two edges or two vertices or an edge and a vertex. In every case the members of a pair must be adjacent in G.

By a c-*colouring* we mean that one of c colours has been given as follows:- *Case* e. To each edge. *Case* v. To each vertex. *Case* ev. To each edge and each vertex. We say a pair is *bad* if both members of the pair get the same colour. We write χ^e, χ^v, χ^{ev} as the case may be for the minimum number c for which there is a c-colouring with no bad pair. Usually χ^e, χ^v, χ^{ev} are written χ, χ', χ'' and called the chromatic numbers of G.

Let Δ be the maximum vertex degree in G. Brooks [2, p.128] proved that $\chi^v \le \Delta + 1$ and found when the inequality is strict. Vizing [2, p.133] proved that $\Delta \le \chi^e \le \Delta + 1$. We say G is class 1 if $\Delta = \chi^e$ and class 2 otherwise. Erdös and Wilson [1] proved that almost all G are class 1. The complete graph K_p is class 1 iff p is even. Trivially $\Delta + 1 \le \chi^{ev}$ and it has long been conjectured that $\chi^{ev} \le \Delta + 2$.

In case e we call a c-colouring *equitable* if at each vertex the numbers of edges of any two colours differ by at most one. We will use a forthcoming theorem of Hilton and de Werra. It says that if c does not divide the degree of any vertex of G then G has an equitable edge c-colouring.

Consider any one of the three cases and let χ denote χ^e, χ^v or χ^{ev} as the case may be. Let $M = M(G, b, c)$ be the maximum number of c-colourings such that each pair is bad in at most b of the c-colourings. Our object is to study M. To avoid trivialities we assume that $1 \le b$ and $2 \le c < \chi$. Clearly M is superadditive, namely

$$M(b_1, c) + M(b_2, c) \le M(b_1 + b_2, c)$$

so $\lim(b \to \infty) M/b$ exists. In particular we are interested in $M(K_p)$ for the complete graph K_p in view of the following observation.

(1) *If H is a subgraph of G then* $M(G) \le M(H)$.

Usually instead of M we write M^e, M^v, M^{ev} as the case may be. For example we obviously have $b \le M^{ev} \le M^e$, $M^v < \infty$.

2. GENERAL RESULTS

We first define two functions $f(c, d)$ and $F(c, d)$. Put $d = d_1 + \ldots + d_c$ where the d_i differ by at most one, then

$$\left. \begin{array}{l} f(c, d) = \binom{d_1}{2} + \ldots + \binom{d_c}{2} \\[6pt] f(c, d) = \binom{d}{2} / f(c,d) \end{array} \right\} \text{ for } 2 \le c < d.$$

Thus for example if d edges of G have a common vertex α then for Case e in any c-colouring there are at least f bad pairs at α. It is easy to see that $c < F$ and

(2) $\lim(d \to \infty) F(c,d) = c$,

(3) $F(c, mc) = F(c, mc -1) = (mc -1)/(m - 1)$ for $2 \le m$.

Consider any case. Let h be a mapping $h:\{1, 2, \ldots, \chi\} \to \{1, 2, \ldots, c\}$. For $1 \le j \le c$ let v_j be the number of i with $h(i) = j$. Assume that the numbers v_j differ by at most one. Let $\mu = \mu(\chi, c)$ be the number of choices for such h. For example if $\chi = mc$ then $\mu = \chi!/(m!)^c$.

Now by the definition of χ there is a χ-colouring of G with no bad pair. Using this fixed colouring, for each of the μ mappings h we change the colouring to a c-colouring in the obvious way. By symmetry each pair is bad in the same number β of the μ derived c-colourings. Thinking about Case e, if we had χ edges with a common vertex α then each h would induce exactly $f(c, \chi)$ bad pairs (of edges) at α, and hence we notice that

(4) $F(x, \chi) = \mu/\beta$.

It now follows that

(5) $c < F(c, \chi) \le \lim(b \to \infty) M/b$.

If P is the total number of pairs in G and B is the minimum number of bad pairs in any c-colouring then clearly

(6) $M/b \le P/B$.

Sometimes this bound is sharp but clearly not always.

3. CASE e. COLOURING EDGES

Now G has a vertex α of degree Δ. In the first c-colouring there are $\lceil \Delta/c \rceil$ edges of the same colour at α, and in the second $\lceil \lceil \Delta/c \rceil /c \rceil$. Repeating this $b + 1$ times we obtain our first result.

THEOREM 1. *If* $c^{b+1} < \Delta$ *then* $M^e = b$.   □

As remarked in Section 2 if $c < \Delta$ there are at least $f(c, \Delta)$ bad pairs (of edges) at α sc

(7) $M^e/b \le F(c, \Delta)$ for $c < \Delta$.

Using (2), (5), (7) we have

$$\lim(\Delta \to \infty)\{\lim(b \to \infty)M/b\} = c \text{ for Cases e and ev}.$$

We can now state the main result of this section.

THEOREM 2. *If* (i) $\Delta = \chi^e$ *or if* (ii) $\chi^e = \Delta + 1 = mc$ *or if* (iii) $G = K_p$ *with* p *odd and* Δ *is not an odd multiple of* c, *then*

$$\text{Lim}(b \to \infty)M^e/b = F(c, \Delta) \text{ for } c < \Delta.$$

Proof. When G is class 1 we are in (i) and the result follows by (5), (7). In (ii) the result follows by (3), (5), (7). In (iii) we have $p = \Delta + 1$. If c does not divide Δ we get an equitable c-colouring of K_p as a special case of the Hilton and de Werra theorem. If 2c divides Δ it is easy to construct one. The result follows by permuting the vertices of this equitable c-coloured K_p in all possible ways. □

Using the last technique and (6) we get

THEOREM 3. $\lim(b \to \infty) M^e(K_p)/b = P/B$. □

Let S denote a class 2 graph with every vertex of degree 3, for instance the Petersen graph T. They call S a *snark*. For c = 2 Theorem 2 (ii) applies, but an elementary argument about two adjacent vertices shows that snarks have equality in (7) for b = 1 and hence for all b. Similarly $M^e(K_3, b, 2) = 3b$ and $M^e(K_5, 1, 2) = 1$ and $M^e(K_5, 2b, 2) = 6b$ and $M^e(T, b, 3) = 15b$. It would be nice to know $M^e(S, b, 3)$. We get strict inequality in (7) for K_7 with c = 2 because $P/B = 105/43 < 5/2 = F(2, 6)$.

4. CASE v. COLOURING VERTICES

Let Q be the largest integer q for which K_q is a subgraph of G. Then corresponding to Theorem 1 we have the following observation.

THEOREM 4. *If* $c^{b+1} < Q$ *then* $M^v = b$. □

Lemma 5. *If* H *is a subgraph of* G *with* $c < \chi(H)$ *then* $M^v \leq b|E(H)|$.

Proof. Suppose the lemma is false, so that we have more than $b|E(H)|$ colourings. If (α, β) is in the set E(H) of edges of H there are at most b colourings in which α and β get the same colour. Hence there is one colouring in which α and β get different colours for every $(\alpha, \beta) \in E(H)$. But this colouring contradicts $c < \chi(H)$. □

The lemma says in particular that if G has a triangle and c = 2 then $M^v \leq 3b$. It also easily yields our next result.

THEOREM 6.   We have $M^V \leq b|E(G)|$ with equality iff $\chi^V(G) = c + 1$ and $\chi^V(G\setminus x) = c$ for all $x \in E(G)$.

By straightforward arguments on the definitions we get

LEMMA 7.   The number $M^V(b, c)$ is the maximum integer m for which there exist $E_1, \ldots, E_m \subset E(G)$ such that

(i) $\quad \chi^V(G\setminus E_i) \leq c$ for $1 \leq i \leq m$,

and

(ii) $\quad$ each edge of G is in at most b of the E_i. □

Next define A as the minimum value of $|E|$ over $E \subset E(G)$ such that $\chi^V(G\setminus E) \leq c$.

THEOREM 8. $M^V/b \leq |E(G)|/A$.

Proof.   In the notation of Lemma 7 we have $A \leq |E_i|$ so $mA \leq \sum |E_i| \leq b|E(G)|$. □

THEOREM 9.   If $c + 1 = \chi^V$ then $A \leq |E(G)|/\binom{c+1}{2}$.

Proof.   Take a $c + 1$ colouring of G with no bad pair. Then put the vertices into the $c + 1$ colour classes. Some two colour classes have at most $|E(G)|/\binom{c+1}{2}$ edges between them, and we choose these edges for A. □

THEOREM 10.   If G is edge transitive then

$$\lim_{b \to \infty} M^V/b = |E(G)|/A.$$

Proof.   Let $E \subset E(G)$ have $|E| = A$ and $\chi^V(G\setminus E) \leq c$. Let the relevant automorphism group of G be $\{\theta_1, \ldots, \theta_m\}$. Let $E_i = E\theta_i$ for $1 \leq i \leq m$ and apply Lemma 7. □

As an example we quote the edge transitive Petersen graph T. Now $|E(T)| = 15$ and $A = 3$ when $c = 2$. It is not too hard to construct five 2-colourings with $b = 1$ so $M^V = 5b$. We could not find a result for vertex transitive graphs. For instance consider the triangular pyramid γ which has 6 vertices and has $M^V(\gamma, b, 2) = 3b$. This small graph γ is in [2, p.171].

Now consider K_p. We recall that the Turán graph has the maximum number of edges among all graphs with p vertices and $\chi^v = c$. Hence for K_p we have $A = f(p, c)$, so by Theorem 10 we have

THEOREM 11. $\lim(b \to \infty) M^v(K_p)/b = F(c, p)$. □

For $c = 2$ on K_5 we found that M^v is 1, 4, 7, 10, according as b is 1, 2, 3, 4. Using (1), (2), (5) and Theorem 11 we see that

(8) $\lim(Q \to \infty)(\lim(b \to \infty) M^v/b) = c$.

Problem. *If G is edge transitive with $\chi = 3$ and every vertex of degree 3 what is A?*

5. CASE ev. COLOURING EDGES AND VERTICES

Looking again at a vertex of degree Δ we see that

$$M^{ev}/b \leq F(c, \Delta + 1) \text{ for } c < \Delta + 1.$$

Using (5) as before we get

THEOREM 12. *If (i) $\Delta + 1 = \chi^{ev}$ or if (ii) $\chi^{ev} = \Delta + 2 = mc$ then*

$$\lim(b \to \infty) M^{ev}/b = F(c, \Delta + 1) \text{ for } c < \Delta + 1. \quad □$$

We have $\chi^{ev}(K_p) = p = \Delta + 1$ so the theorem covers K_p.

REFERENCES

1. P. Erdös and R.J. Wilson, On the chromatic index of almost all graphs, J. Combinatorial Theory Series B, <u>23</u> (1977) 255-257.

2. F. Harary, *Graph Theory*, Addison-Wesley (1969).

ON HADWIGER'S NUMBER AND THE STABILITY NUMBER

P. DUCHET and H. MEYNIEL

E.R. 175, C.M.S.
54 Bd. Raspail, 75007 Paris, France

If a graph with n vertices and stability number α is not contractible to K_{h+1} then the following inequality holds:

$$(2\alpha - 1) h \geq n.$$

Let G be a graph (multiple edges or loops are permitted); we use here the following parameters:

$n(G)$: the number of vertices
$\alpha(G)$: the stability number
$\gamma(G)$: the chromatic number
$h(G)$: the Hadwiger number (= the greatest integer h such that G admits the complete graph K_h as subcontraction)
$\beta(G)$: the minimal cardinality of a dominating set of G.

For every graph we have

(1) $\alpha \gamma \geq n.$

As a special case of the four-colour problem, Erdös had conjectured that for every planar graph the following inequality holds:

(2) $\alpha \geq \frac{n}{4}.$

This is actually implied by the four-color theorem [3]. Thus, as a consequence of Hadwiger's conjecture [1]:

HADWIGER'S CONJECTURE : $h \geq \gamma$

a very natural question appears, namely

CONJECTURE : $\alpha \geq \frac{n}{h}$.

In [2], Mader proved that a graph which is not contractible to K_p has at most $8n \lceil p \log_3 p \rceil$ edges. Therefore we have

(3) $\gamma \leq 16 \lceil (h+1)(\log_3 (h+1) \rceil$.

Hence

(4) $\quad \alpha \geq \dfrac{n}{16\lceil (h+1)\log_3(h+1)\rceil}$.

Our aim here is to improve inequality (4) as follows.

THEOREM. For every graph,

(5) $\quad (2\alpha-1)\cdot h \leq n$.

LEMMA. For every connected graph, there exists a connected dominating subset with β' vertices such that

(6) $\quad \beta' \leq \min(2\alpha-1, 3\beta-2)$.

PROOF OF THE LEMMA. We construct by induction two sequences of sets of vertices (S_i) and (D_i) with the following properties:

$S_0 \subset S_1 \subset S_2 \ldots$ are stable sets,

$D_0 \subset D_1 \subset D_2 \ldots$ are connected sets,

$S_i \subset D_i$ and $|D_i| \leq 2|S_i| - 1$ for $i = 0,1,2,\ldots$.

We begin with $S_0 = D_0 = \{v\}$ where v is any vertex of G. If S_i is not a maximal independent set, we can choose a vertex x at distance two from S_i since the graph is connected. Let y be a common neighbour of x and S_i and put

$S_{i+1} = S_i \cup \{x\}$,

$D_{i+1} = D_i \cup \{x,y\}$.

The last S_i is a maximal independent set and D_k is the required connected dominating set.

In order to prove the inequality $\beta' \leq 3\beta - 2$ an analogous procedure is possible. We construct sequences (C_i) and (D_i) with the properties:

$C_0 \subset C_1 \subset C_2 \ldots$ are connected sets,

$D_0 \subset D_1 \subset D_2 \ldots$ are subsets of D,

$D_i \subset C_i$ and $|C_i| \leq 3|D_i| - 2$ for $i = 0,1,2,\ldots$,

where D is a given dominating set of the graph which is minimal for inclusion.

We start again with $C_0 = D_0 = \{v\}$, where v is any vertex of D. If $D_i \neq D$, then, by the minimality of D, there exists a vertex y at distance two from D_i; hence D contains a vertex x at distance at most 3 from D_i (x is y or a neighbour of y). Thus we put

$$D_{i+1} = D_i \cup \{x\},$$
$$C_{i+1} = C_i \cup \{x,y,z\},$$

where z is a common neighbour of y and D_i.

When $D_k = D$, C_k is the required connected dominating set.

REMARK. In (6), the inequalities are the best possible between β' and β or α. Nevertheless, a slight refinement of $\beta' \leq 2\alpha - 1$ can be shown: a connected graph always contains a connected dominating subset with at most $2\alpha - 1$ vertices that contains a stable subset with α vertices. The proof is easy by induction on the number of vertices.

PROOF OF THE THEOREM. Put $G = (V, E)$. Clearly we can suppose that G is connected. If D is a connected dominating subset, then by considering the effect of the contraction by D we find that

$$h(G_{V-D}) \leq h(G) - 1,$$

where G_{V-D} denotes the induced subgraph.

Since we can choose D with no more than $2\alpha - 1$ vertices, the inequality (5) easily follows by induction on $n(G)$.

REFERENCES

[1] H. Hadwiger, Ungelöste probleme, Element. Math. 13 (1958).

[2] W. Mader, Homomorphiesatze für Graphen, Math. Annalen 178 (1968) 154-68.

[3] K. Appel and W. Haken, Every planar map is 4-colourable, 1 & 2 Illinois J. Math. 21 (1979), 429-490 and 491-567.

A DEPTH-FIRST-SEARCH CHARACTERIZATION OF PLANARITY

H. DE FRAYSSEIX and P. ROSENSTIEHL

Collège de France, Paris, France
and
Ecole des Hautes Etudes en Sciences Sociales, Paris, France

1. INTRODUCTION

Hopcraft and Tarjan [1] have produced the first algorithm for testing planarity of graphs in linear time. Their proof of linearity relies on a lemma associated with the computing process.

The main theorem of this paper displays a property of the Trémaux trees of a graph, which constitutes a new characterization of planarity, and justifies the so-called "Left-Right Algorithm for Planarity Testing and Embedding" [6]. In some way the theorem enlightens in general terms why a Trémaux tree allows a linear computing time. A theorem on Trémaux trees is also a theorem on Depth-First-Search [4], since the trees of the maze solver - le Sir de Trémaux - are exactly the trees generated by Depth-First-Search.

Let us start with some definitions. We consider as well known the usual definitions of graph, vertex, edge, cycle, cocycle and tree. By a tree we shall mean a spanning tree. Any rooted tree (T,r) of G defines a partial order on the vertices of G, for which r is the minimum element. A *Trémaux tree* (or *Depth-first-search tree*) of G is a rooted tree (T,r) such that each cotree edge $\alpha \in T^{\perp}$ is incident with two comparable vertices. It follows that every edge of G, belonging to T or $T^{\perp}$, is incident with a lower vertex $v^{-}(a)$ and an upper vertex $v^{+}(a)$, and we have

$$r \leq v^{-}(a) < v^{+}(a) .$$

The *fundamental cycle* $T(\alpha)$, defined for $\alpha \in T^{\perp}$, is the cycle which meets $T^{\perp}$ in α. The *fundamental cocycle* $T^{\perp}(e)$, defined for $e \in T$, is the cocycle which meets T in e. A *crossable pair* $[\alpha, \beta]$ is a pair of edges $\alpha, \beta \in T^{\perp}$ incident with four different vertices, and such that $T(\alpha) \cap T(\beta)$ is not empty. A *Trémaux embedding* G^{O} of G is an embedding such that the pairs of edges which cross have only one crossing and are crossable pairs.

The *crossing set* X^o is the set of crossable pairs that cross in G^o. To a crossable pair $[\alpha,\beta]$ whose upper vertices are not comparable, correspond a unique pair of tree edges, e and f, such that $v^-(e) = v^-(f)$ and $\alpha \in T^\perp(e)$, $\beta \in T^\perp(f)$. The switching set $\mu(e,f)$ associated with such a crossable pair $[\alpha,\beta]$ is the set of crossable pairs included in the cartesian product of $T^\perp(e)$ and $T^\perp(f)$. Geometrically, two Trémaux embeddings G^o and G' which differ by switching of e and f only, have crossing sets X^o and X' whose symmetrical difference is equal to $\mu(e,f)$ (Figure 1).

Let us note that an edge $\alpha \in T^\perp$ defines an upper angle $\theta^+(\alpha)$ and a lower angle $\theta^-(\alpha)$ with the tree chain which joins $v^+(\alpha)$ to $v^-(\alpha)$. In an embedding G^o, an angle is either on the right or on the left of the corresponding chain.

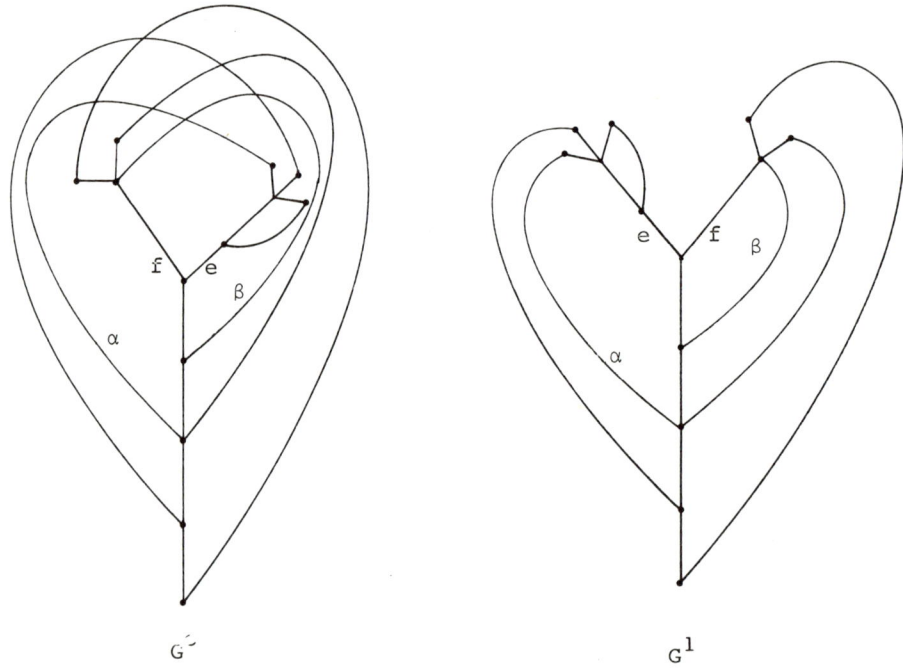

Figure 1 - Switching e and f.
G satisfies the conditions of the theorem.
G^o is compatible with the cocycle D.
$X^o = \mu(e,f)$. $X' = 0$. G is planar.

(Straight lines for tree and curved lines for cotree edges.)

Let us give now a theorem on which relies the main theorem of the paper.

Theorem 1. *If the crossing set X^o of a Trémaux embedding G^o of G can be partitioned into switching sets $\mu(e,f)$, where e and f are tree edges with a common lower vertex, then G is planar.*

The Wu-Liu characterization theorem of planar graphs [3] defines a vector space Σ on $GF(2)$ generated by three kinds of switching sets, including those of the $\mu(e,f)$ type, and asserts that G is planar if and only if $X^o \in \Sigma$. Thus it includes Theorem 1.

2. THE MAIN THEOREM

Let us recall that two angles are defined for each cotree edge of a Trémaux tree of G, and that in a given embedding some pairs of angles are on the same side, some others are on different sides.

We consider below a graph G, a Trémaux tree (r,T) of G, and five types of pairs of angles in order to define two abstract binary relations on angles. The first class of pairs will be called S (S for "same") and a second class D (D for "different"). S ∪ D will constitute the edges of a graph H whose vertices are the set of cotree angles. α, α' and β denote cotree edges.

Type (i): $[\theta^+(\alpha), \theta^-(\alpha)] \in S$
whenever there exists $\beta \in T^\perp$ such that
$v^-(\beta) < v^-(\alpha) < v^+(\alpha) < v^+(\beta)$.

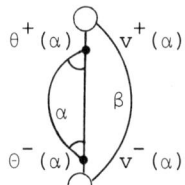

Type (ii): $[\theta^+(\alpha), \theta^-(\beta)] \in D$ if
$v^-(\alpha) < v^-(\beta) < v^+(\alpha) < v^+(\beta)$.

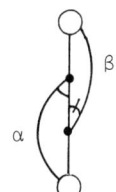

Type (iii): $[\bar{\theta}(\alpha), \bar{\theta}(\beta)] \in D$
if there exist $e, f \in T$ and
$\beta' \in T^\perp$ such that
$\alpha \in T^\perp(e)$ and $\beta, \beta' \in T^\perp(f)$ with
$\bar{v}(\beta') < \bar{v}(\alpha) < \bar{v}(\beta) < \bar{v}(e) = \bar{v}(f)$.

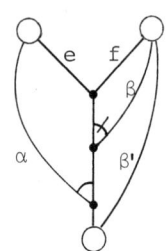

Type (iv): $[\bar{\theta}(\alpha), \bar{\theta}(\beta)] \in D$
if there exist $e, f \in T$ and
$\alpha', \beta' \in T^\perp$ such that
$\alpha, \alpha' \in T^\perp(e)$ and $\beta, \beta' \in T^\perp(f)$ with
$\bar{v}(\alpha') = \bar{v}(\beta') < \bar{v}(\alpha) = \bar{v}(\beta) < \bar{v}(e) = \bar{v}(f)$.

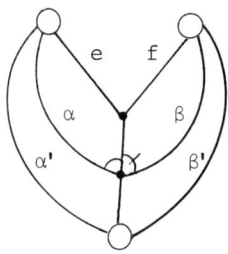

Type (v): $[\bar{\theta}(\alpha), \bar{\theta}(\alpha')] \in S$
if there exists $e, f \in T$ and $\beta \in T^\perp$
such that
$\alpha, \alpha' \in T^\perp(e)$, $\beta \in T^\perp(f)$ with
$\bar{v}(\beta) < \bar{v}(\alpha) \le \bar{v}(\alpha') < \bar{v}(e) = \bar{v}(f)$.

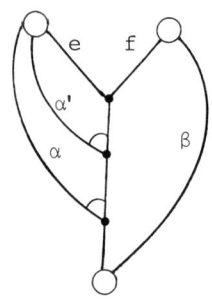

(In the figures, circles are used to indicate that location and side are not considered.)

When G^0 is an embedding without crossing, it is easy to check that the set D (resp. S) defined above in an abstract way is a subset of pairs of cotree angles disposed on different sides (resp. same side) in the embedding.

The main theorem considers in some way the reciprocality.

Theorem 2. G *is planar if and only if the set* D *is a cocycle of* H.

Outline of the proof

I. Let G^o be a planar embedding without crossing. It defines for each angle a side : left or right.

It is easy to check for each of the five types that, whenever a pair of angles belongs to D (resp. S), the angles are arranged on different sides (resp. same side) in the embedding, for if they were not, crossings would appear. Therefore D is a cocycle of H, which separates the left angles from the right ones.

II. Let G be a graph such that, for a given Trémaux tree, its associated set D is a cocycle of H. First it is easy to construct a planar Trémaux embedding G^o (with possible crossings) such that, whenever a pair of angles belongs to D (resp. S) the angles are arranged on different sides (resp. same side). In other words G^o satisfies the five types of relations.

By Theorem 1 the proof will be complete if we show that the crossing set X^o can be partitioned into switching sets $\mu(e,f)$.

Let $[\alpha,\beta] \in X^o$. From conditions (i) and (ii), $v^+(\alpha)$ and $v^+(\beta)$ are not comparable. Therefore there exists a unique pair $e,f \in T^\perp$ associated to α and β. From conditions (iii), (iv) and (v) it follows that $\mu(e,f) \subset X^o$; in other words, for any crossable pair $[\alpha',\beta']$ with $\alpha' \in T^\perp(e)$, $\beta' \in T^\perp(f)$, we have $[\alpha',\beta'] \in X^o$ (see Figure 1).

3. CONCLUSION

In the context of Depth-First-Search, planarity testing and embedding request to deal only with relations on cotree angles, defined by order relations as displayed in cases (i),(ii),(iii),(iv), (v). In particular the embedding of the tree can be ignored, since tree angles do not appear in the necessary conditions of Theorem 2.

Because of the drastic reduction of variables to deal with, the linear computing time can be achieved.

REFERENCES

[1] Hopcroft, J.E., and Tarjan, R.E., Efficient planarity testing, J. ACM, 21 (1974), 549-568.
[2] Liu, Y., *Acta Math. Appl. Sinica* 1 (1978), 321-329.
[3] Rosenstiehl, P., Preuve algébrique du critère de planarité de Wu-Liu, in *Annals of Discrete Mathematics*, 9, (1980), 67-78.
[4] Tarjan, R.E., Depth-first-search and linear graph algorithm, *SIAM J. Comput.* (2), (1972), 146-160.
[5] Tutte, W.T., Toward a theory of crossing numbers, *J. Combinatorial Theory*, 8, (1970), 45-53.
[6] Fraysseix, H. de, and Rosenstiehl, P., The left-right algorithm for planarity testing and embedding in linear time (to appear)

GRAPH-THEORETICAL MODEL OF SOCIAL ORGANIZATION*

JONATHAN L. GROSS

Department of Computer Science
Columbia University
New York, N.Y. 10027, U.S.A.

Mary Douglas has proposed a theory of cultural analysis that focuses on two mechanisms for the control of social behavior. One is the totality of categorical distinctions, such as day and night. The other is the extent to which individuals experience their own identities through membership in social units. What distinguishes this theory from most others are the fundamental hypotheses that these two mechanisms are quantifiable and that they are operationally orthogonal. Anticipating their precise formulation, she has introduced the names <u>grid</u> and <u>group</u> for measured values of the total strength of these two respective mechanisms. Polytechnic models of grid and group have been developed and are described here. The outcome of an experiment partially confirming the reliability of grid measurements is also presented.

1. INTRODUCTION

Certain sociological phenomena are apparently better understood through analysis of a cultural environment than of individual behavior. The classic case is suicide, as reported by E. Durkheim [1897]. Another striking example is the success of S. Minuchen in treating anorexia by modifying the structure of the family relationships, rather than by isolating the patient with the presenting symptom.

Some of the phenomena of interest in this program of research are competitiveness, bureaucratic standardization, and identification

---

* Dedicated to Professor Frank Harary on the occasion of his sixtieth birthday.

of individual interest with collective goals. Mary Douglas has
identified the various kinds of cultural environments in which each of
these attributes is likely to exist. For the past three years I have
been developing mathematical models of social structure and designing
experiments to test her cultural insights. The preliminary empirical
results are positive.

I have been able to quantify what Douglas [1970, 1978] considers
to be the two first-order variables of cultural environment. One,
called <u>grid</u>, is the amount of control exerted by the publicly
acknowledged categorical distinctions such as male and female,
workday and weekend, or edible and inedible. The measured value
depends only on the extent to which such classifications control
behavior in that environment, not on what particular categories exist
or how they are observed. The other, called <u>group</u>, first-order
variable is the amount of control that arises from a person's
experiencing his or her identity through membership in a social unit.
For instance, a wage-earner who voluntarily purchases life insurance
and names his wife and children as beneficiaries might be behaving
in accordance with his culturally consistent experience of their
needs as his own.

The quantification of grid and group is based on a graph-theoretic
model, called the EXACT model, which I developed in collaboration
with Steve Rayner. Once a social unit is represented in terms of an
EXACT model, it is possible to measure its grid score and its group
score. Both of these are derived from empirical data. An experiment
providing evidence of the reliability of one of the predicates is
described.

This brief sketch indicates the basic feature of a combinatorial
model of social organization. Details of the empirical methodology
and justification of the choice of predicates are presented in a
monograph by Gross and Rayner [1982].

For a general source on graph theory, the reader is referred to
F. Harary [1969].

2. THE EXACT MODEL

The EXACT model for a social unit has the following components:

- X = the set of members of the unit
- A = the set of defining activities of the unit
- C = the set of roles persons assume in these activities
- T = a cultural partition of the annual time cycle
- E = the set of eligible nonmembers

Thus, "EXACT" is an approximate acronym.

Although $X = \{x_1, \ldots, x_p\}$, $A = \{a_1, \ldots, a_m\}$ and $E = \{e_1, \ldots, e_n\}$ are simply sets, both C and T have additional structure. In particular, there is a subset of roles.

$$_iC = \{_iC_1, \ldots, _iC_{i_r}\}$$

associated with activity a_i. The timespan is partitioned into culturally-defined time units such as months, weeks, and holidays. It also carries the structure of life-cycle events and time-factors in social reciprocity. In precise terms, this structure is representable as probabilistic distributions and algorithms.

A multigraph associated with this model is called the EXACT graph. Its vertices are the members of the unit. For every activity a_i and every pair of members x_j and x_k who interact in activity a_i, there is an edge labeled a_i with endpoints x_j and x_k. In a more elaborate version of the EXACT model, this edge would also carry a weighting label to indicate the extent of the interaction.

As a social unit, one might theoretically select any collection of persons such that the EXACT graph is connected. In practice, one aims for a collection of persons such that, subject to various constraints, the grid and group scores are locally maximal. In anthropological jargon, one would say that our social units are defined by the culture. Alternative bases for defining social units might be geographic (e.g. all the persons in village Y) or political (e.g. every card-carrying member of organization Z). Often these criteria might yield the same selection of a social unit. However, in cases of juxtaposed cultures, they yield different units.

3. THE GROUP PREDICATES

At present, the group score is a composite of five "predicate" scores, each associated with the EXACT graph G. Each predicate score corresponds to a particular social attribute and is a number between 0 and 1.

The first of the group predicates is <u>proximity</u>, which is the inverse of distance. Of course the distance $d(x_i, x_j)$ between the two vertices x_i and x_j is the minimum number of edges one must traverse to get from x_i to x_j. Thus, the average distance from x_i to the other vertices is a positive integer. It follows that the reciprocal of the average distance is a number between 0 and 1, which we denote by $prox(x_i)$. Then $prox(G)$ is defined to be the average vertex proximity.

The second group predicate is <u>transitivity</u>. From the perspective of x_i, the issue is whether when two persons x_j and x_r both interact with x_i they also interact with each other. Let N_i denote the induced (simplicial) graph on the neighbors of x_i. Then
$$trans(x_i) = \#E(N_i) / \binom{\#V(N_i)}{2}$$

Of course, $\#E(N_i)$ and $\#V(N_i)$ mean the numbers of edges and vertices of N_i, respectively. We now define $trans(G)$ to be the average vertex transitivity.

As graph invariants, there is no reason why high proximity and low transitivity cannot coexist. For instance a complete bipartite graph $K_{p,p}$ has proximity at least 2/3 but transitivity 0. On the other hand, imagine a graph constructed by running a long edge path P_u from one complete graph K_v to another copy of K_v. Then the proximity is approximately $2/u$. If the number v is much larger than the number u, then the transitivity score is nearly 1. The distribution of pairs of values $(prox(G), trans(G))$ for actual social units is a matter for experimental study.

The third group predicate is <u>frequency</u>. We define $freq(x_i)$ to be the proportion of his or her allocable time that x_i interacts with the other members of the unit. Also, $freq(G)$ is the average of the individual frequencies. The obvious alternatives to interacting with other members of the unit are interacting with outsiders or with no one.

If a given social unit, such as family, is a subunit of a larger system, or if its membership overlaps with other social units, then a person's commitment might extent beyond the given unit. The fourth predicate score, scope, measures the diversity of a person's involvement in activities of the unit relative to total activity. More precisely, scope (x_i) is the ratio of the number of intra-unit activities in which x_i interacts with other members to the total number of activity inside and outside the unit in which x_i interacts with other persons. Scope (G) is the average individual scope.

The fifth group predicate score, impermeability, is the only one that has no local values. During the annual time cycle some subset of eligible non-members become members of the unit. We define

$$\text{imperm (G)} = 1 - \frac{\#\text{new members}}{\#\text{eligibles}}.$$

At present, we use the average of the five group predicates for the score group (G). It is possible that additional empirical evidence will motive the adoption of a weighted average. If for some reason, not all five predicates are applicable to some social unit, one averages the applicable ones.

The predicate scores appear to be calculable only if one watches everyone on the unit for a year. Obviously one uses samples to obtain approximate scores.

4. THE GRID PREDICATES

Like group, grid is a composite of up to five predicates, each scored between 0 and 1. A predicate score for an entire social unit is calculated as an average over all members of their scores for that predicate.

The first grid predicate is specialization, the opposite of diversification. Let $C(x_i)$ be the set of roles assumed during the time cycle T. Then $\text{spec}(X) = 1 - \#C(x_i)/\#C$. Thus, a person who assumes very few roles will score near 1.

The second grid predicate is <u>asymmetry</u>. For each pair of members x_i and x_j we define the interaction number $\text{inter}(x_i, x_j)$ to be the number of activities in which x_i and x_j interact as a consequence of their roles. Also, we define the exchange number $\text{exch}(x_i, x_j)$ to be the number of activities in which x_i and x_j exchange roles and interact from both sides. For instance, in a hospitality activity in many cultures, it is likely that x_i and x_j exchange the roles of host and guest. In an employment activity it is unlikely that the roles of boss and worker are exchanged. We define the asymmetry score exchanged.
We define the asymmetry score

$$\text{asym}(x_i) = 1 - \frac{1}{\#V(N_i)} \sum_{x_j \in V(N_i)} \frac{\text{exch}(x_i, x_j)}{\text{inter}(x_i, x_j)}$$

The third grid predicate is <u>entitlement</u>. We calculate the score for a member x_i as the proportion of his or her roles that are attained by ascription, rather than by achievement. For instance, if someone is automatically a carpenter because his father was one, that is ascription. On the other hand, if he gets a job because he is so skilled at it, that is achievement. In general, we define

$$\text{entitle}(x\frac{1}{c}) = \frac{\#\text{ascribed roles}(x_i)}{\#\text{roles}(x_i)}$$

The fourth grid predicate is <u>accountability</u> for one's immediate behavior. The scoring is based on situations in which the occupants of two roles interact. In an "accountability interaction", one role (the "dominant" role) includes sanctions to be used in exacting performance from the person in the other role (the "regulated" role). For each member x_i we define the accountability score $\text{acc}(x_i)$ to be the proportion of his interactions that are accountability interactions.

The fifth grid predicate is <u>information</u> transmission. Phenomena such as food-taking, the wearing of clothing, or the allocation of space may be regarded as behavioral media for the transmission of information about time and social roles. The general idea is there exists optimal strategies for guessing what roles or events are being represented both in the absence and in the presence of clues from such media. From the probabilities of correct guesses with and without the clues, one can calculate the information value of the clues, in the sense of C. Shannon [1948]. The information score info(x_i) is based on the value of the clues provided by the behavior of x_i.

5. AN EXPERIMENT: INFORMATION IN FOOD BEHAVIOR

The worth of their quantification of grid and group depends on two things, the reliability of the empirical measurements and their use in the analysis of such social traits as competitiveness, bureaucratic standardization, and consciousness of group welfare. An experimental program sponsored by the Russell Sage Foundation under the direction of Mary Douglas has provided evidence to support the reliability of the scoring method for one of the grid predicates, information transmission.

The method was applied in four to eight households in each of three different American communities: a southern rural community, an ethnic suburb of a northeastern city, and a midwestern Indian reservation. In each household of each community, the clues to time-based events were five or more aspects of food-taking behavior, including the duration of the meals, the number of persons attending and the number of different food varieties served. For each aspect and each household an information score was calculated. Due to the small size of the sample, only the ranks were considered significant.

We discovered that from one aspect to another, the information rankings were consistent. This tends to confirm the reliability of the method. What it suggests is that persons internalize consistent syndromes of behavior in accordance with Douglas's theory of cultural analysis. Full details are presented by Gross [1981].

REFERENCES

[1]   M. Douglas [1970], <u>Natural Symbols</u>, Penguin.

[2]   M. Douglas [1978], <u>Cultural Bias</u>, Royal Anthropological Institute, London.

[3]   E. Durkheim [1897], <u>Le Suicide</u>, translated into English as <u>Suicide</u>, Rout Ledge & Kegan Paul, London, 1952.

[4]   J.L. Gross [1981], Measurement of calendrical information in food-taking behavior, manuscript.

[5]   J.L. Gross and S.F. Rayner [1982], Grid/group analysis of social organization: an operational guide, manuscript.

[6]   F. Harary [1969], <u>Graph Theory</u>, Addison-Wesley.

[7]   C.E. Shannon [1948], A mathematical theory of communication, <u>Bell System Technical Journal</u> 27.

ODD CYCLES OF SPECIFIED LENGTH IN NON-BIPARTITE GRAPHS

ROLAND HÄGGKVIST

Institut Mittag-Leffler
Djursholm, Sweden

It is proved that every sufficiently large graph with minimum degree more than $\frac{2n}{2k+1}$ either contains no odd cycle of length at least $\frac{k+1}{2}$ or else it contains a $(2k-1)$-cycle.

It is also shown that every triangle-free graph with minimum degree more than $\frac{3n}{8}$ is homomorphic with a 5-cycle.

At the end some conjectures are formulated.

1. INTRODUCTION

For unexplained terminology see [2]. In particular note that C_m denotes the cycle of length m, that $\delta(G)$ denotes the minimum vertex-degree in the graph G, and that G is H-free if G has no subgraph isomorphic with the given graph H. In what follows the number n will always stand for the number of vertices in the graph G under consideration.

One of the best-known results in graph theory reads: Every graph with at least $\frac{n^2}{4}$ edges contains a triangle, unless it is bipartite. This special case of Turán's theorem [8] was proved by Mantel in 1907 [7], and has been the starting point of many investigations, See Bollobás [3, Ch III.5 and ch VI] for a representative sample of the results which have been obtained.

Here I only note that Erdös (together with Gallai), and independently Andrásfai, proved that a triangle-free non-bipartite graph has at most $\left[\frac{(n-1)^2}{4}\right] + 1$ edges [5, Lemma 1]; the same proof gives under the same assumptions that the minimum degree is at most $\frac{2n}{5}$, and in fact, as noted in Andrásfai, Erdös and Sós [1, remark 1.6], that every non-bipartite graph with minimum degree greater than $\frac{2n}{2k+1}$ contains an odd cycle of length at most $2k-1$.

In this note I shall present some results on C_{2k-1}-free graphs. It will be shown that if such a graph contains an odd cycle of length at least $\frac{k+1}{2}$ then its minimum degree is at most $\frac{2n}{2k+1}$ if n is sufficiently large with respect to k. Hence, any 2-connected non-bipartite sufficiently large graph with minimum degree greater than $\frac{2n}{2k+1}$ contains a cycle of length $2k-1$. This is sharp in view of the graphs $C_{2k+1}(m)$, where $H(m)$ denotes the graph obtained from H by replacing each $x \in V(H)$ by an m-tuple, joining every vertex in X to every vertex in Y iff (x,y) is an edge in H.

The assumption that G is 2-connected can be dropped if $k = 2,3,4$ or 5, but is in general necessary.

I shall also demonstrate that every triangle-free graph with minimum degree greater than $\frac{3n}{8}$ is homomorphic with C_5 (i.e. $G \subset C_5(n)$). This result should be compared to that of Andrásfai, Erdős and Sós [1] which (reformulated) states: Every K_r-free graph with minimum degree greater than $\frac{3r-7}{3r-4}n$ is homomorphic with K_{r-1} (i.e. has chromatic number less than r). Restricted to the case $r = 3$ this again gives that any triangle-free graph with minimum degree greater than $\frac{2n}{5}$ is bipartite.

Finally, at the end some conjectures are stated.

2. MAIN RESULTS

Let me first give a proof of the result of Erdős, Gallai, Andrásfai and Sós mentioned above.

Theorem 1. Every non-bipartite graph with $\delta > [\frac{2n}{2k+1}]$ contains an odd cycle of length at most $2k-1$.

Proof. Let S be a shortest odd cycle in G. We may assume that $\nu(S) \geq 5$ since else the conclusion in the theorem holds trivially. S is clearly chord-free, and no vertex outside S is joined to S by more than two edges, since S is selected as a shortest odd cycle in G. Let $E(A,B)$ denote the set of edges in G with one end in A and the other in B. Then we have

$$\sum_{x \in V(S)} d(x) = \sum_{x \in V(S)} |E(x,S)| + \sum_{x \in V(S)} |E(x,G-S)|$$

$$= \sum_{x \in V(S)} |E(x,S)| + \sum_{y \in V(G)-V(S)} |E(S,y)|$$

$$\leq 2\nu(S) + 2(\nu(G)-\nu(S))$$
$$= 2\nu(G)$$

Hence $\sum_{x \in V(S)} d(x) \leq 2\nu(G)$ which in particular implies that $\delta(G) \leq [\frac{2n}{2k+1}]$ if S has length at least $2k+1$. This proves theorem 1.

Remark. The above proof gives immediately that if S is a shortest odd cycle in a triangle-free graph G then $\sum_{x \in V(S)} d(x) \leq 2n$. In particular a non-bipartite graph G which satisfies any of i/-iii/ below contains an odd cycle of length at most $2k-1$.

i/ $d(x)+d(y) > \frac{4n}{2k+1}$ for every pair of non-adjacent vertices x,y in G.

ii/ $d(x)+d(y) > \frac{4n}{2k+1}$ for every pair of adjacent vertices x,y in G.

iii/ $\sum_{x \in V(P)} d(x) > \frac{2(t+1)n}{2k+1}$ for every induced path of fixed length t, $t \leq 2k$.

Theorem 2. Let G be a graph with $\delta > [\frac{2n}{2k+1}]$ and $n > \binom{k+1}{2}(2k+1)(3k-1)$. Then, either G contains a C_{2k-1} or else G contains no odd cycle of length greater than $\frac{k}{2}$.

Theorem 2 has some immediate consequences. By a theorem of Voss and Zuluaga [9] every 2-connected non-bipartite graph contains an odd cycle of length min $\{2\delta-1,n\}$. Consequently we have

Corollary 1. Let G be a 2-connected non-bipartite graph with $\delta > [\frac{2n}{2k+1}]$ and $n > \binom{k+1}{2}(2k+1)(3k-1)$. Then G contains a C_{2k-1}. Also note that since every non-bipartite graph contains an odd cycle (of length at least three) we have

Corollary 2. Let G be a non-bipartite graph with $\delta > [\frac{2n}{2k+1}]$ and $n > \binom{k+1}{2}(2k+1)(3k-1)$. Then, if k = 2,3,4 or 5, G contains a C_{2k-1}.

Corollary 1 is best possible because of the family of graphs $C_{2k+1}(m)$, and Corollary 2 is also sharp (Take e.g. three copies of $K_{m,m}$, select a vertex in each copy and join the chosen vertices by altogether three edges. The resulting graph has 6m vertices, minimum degree m, is clearly non-bipartite, and has no 11-cycle.)

Proof of theorem 2. First some lemmas.

Lemma 1. Let A_i, $i = 1,2,\ldots m$, be sets. Then

$$\left| \bigcup_{i=1}^{m} A_i \right| \geq \sum_{i=1}^{m} |A_i| - \sum_{1 \leq i < j \leq m} |A_i \cap A_j|.$$

Proof. This is nothing else but the first two terms in the inclusion-exclusion formula.

Lemma 2. Let S be a set of $k+1$ vertices in G (the graph in Theorem 2). Then S contains a pair of vertices u,v for which $|N(u) \cap N(v)| > 3k-2$.

Proof. By lemma 1 $\left| \bigcup_{v \in S} N(v) \right| + \sum_{\substack{u \neq v \\ u,v \in S}} |N(u) \cap N(v)| \geq \sum_{v \in S} |N(v)|.$
Hence

$$n + \sum_{\substack{u \neq v \\ u,v \in S}} |N(u) \cap N(v)| > (k+1) \left[\frac{2n}{2k+1}\right] \geq (k+1)\left(\frac{2n}{2k+1} - \frac{2k}{2k+1}\right).$$

This implies $\sum_{\substack{u \neq v \\ u,v \in S}} |N(u) \cap N(v)| > \frac{n}{2k+1} - \frac{2k^2+2k}{2k+1} > \binom{k+1}{2}(3k-1) - 2k-1.$

Since the sum has $\binom{k+1}{2}$ summands, at least one of these must be greater than

$$3k-1 - \frac{2k+1}{\binom{k+1}{2}} > 3k-2, \quad \text{since} \quad k > 1.$$

This proves the lemma.

Definition. An (m,k)-starter in G is a set of $k+1$ vertices, each pair of which are joined by a path of length $2j-1$, for some $j \in \{m, m+1, \ldots k-1\}$.

Lemma 3. If G contains an (i,k)-starter for some $i < k$, then G contains a C_{2k-1}.

Proof. Let S be an (i,k)-starter in G, with i maximum. By lemma 2 S contains a pair of vertices x,y such that $|N(x) \cap N(y)| > 3k-2$. By definition of an (i,k)-starter, x is joined to y by a path $p: x-x_1-\ldots x_{2j-2}-y$ of length $2j-1$ for some $j \in \{i, i+1, \ldots k-1\}$. We consider two cases.

Case 1. $j = k-1$. In this case let z be a vertex in $N(x) \cap N(y) - V(P)$. We know that z exists since $3k-2 > 2j$. The cycle $C: z-x-x_1-\ldots x_{2j-2}-y-z$ is a cycle of length $2k-1$, which proves the lemma in the first case.

Case 2. $j < k-1$. Then $|N(x) \cap N(y) - V(P)| > 3k-2-(2k-4) > k+2$. Consequently $N(x) \cap N(y) - V(P)$ contains a set Q of $k+1$ vertices, any pair u,v of which are joined by a path $u-x-x_1-\ldots-x_{2j-2}-y-v$ of length $2j+1$. It follows that Q is a $(j+1,k)$-starter, contrary to the choice of S (since $i < j+1 \leq k-1$). This contradiction proves lemma 3.

Lemma 4. If G contains an odd cycle of length at least $\frac{k+1}{2}$, then G contains a $(1,k)$-starter.

Proof. We say that an odd cycle of length at least $\frac{k+1}{2}$ is admissible and choose $C: c_1-c_2-\ldots-c_{2m-1}-c_1$ to be a shortest admissible cycle in G. We examine five cases.

Case 1. $2m-1 < k+1$. Since by assumption $\delta(G) > [\frac{2n}{2k+1}] > 2(k+1)$ there exist distinct vertices u_i, $i = 1,2,\ldots 2m-1$, in $G-V(C)$ such that (C_i, u_i) is an edge in G for every i. The set $V(C) \cup \{u_1, u_2, \ldots u_{k+2-2m}\}$ is a $(1-k)$-starter in G.

Case 2. $k+1 < 2m-1 < 2k-1$. In this case $\{c_1, c_2, \ldots c_{k+1}\}$ is a $(1,k)$-starter.

Case 3. $2k-1 < 2m-1$ and C has a chord (c_i, c_j). Let P_{c_i, c_j} be the segment of even length between c_i and c_j along C. Then the length of P_{c_i, c_j} is no more than $\frac{k}{2} - 1$ at most, since else C fails to be a shortest admissible cycle. We may therefore assume that $c_i = c_1$ and $c_j = c_{2r-1}$ for some r such that $2r-1 < \frac{k}{2}$. Put $S_1 = \{c_1, c_2 \ldots c_{2r-1}\}$. Now, if $m-r$ is odd, say $m-r = 2q+1$, then we put

$S_2 = \{c_i : i = 2r-1+4t-1$ for $t = 1,2,\ldots q\}$

and if $m-r$ is even, say $m-r = 2q$, then we put

$S_2 = \{c_i : i = 2r-1+4p-3$ for $p = 1,2,\ldots q\}$.

In both cases put

$S_3 = \{c_i : i = 2r-1+4p$ for $p = 1,2,\ldots q\}$.

Then $S = S_1 \cup S_2 \cup S_3$ is a set of at least $k+1$ vertices (equality holds when $(r,m) = (2,k+1)$) any pair of which are joined by an odd path of length at most $2m-5$, using vertices from C only. Indeed, this is obviously true for any pair of vertices in $S_2 \cup S_3$ since no such pair of vertices are of distance 2 along C. Any pair of vertices in S_1 are joined by an odd path in $P_{c_1,c_{2r-1}} + (c_1,c_{2r-1})$ of length at most $2r-3$ and clearly if $x \in S_1$ and $y \in S_2 \cup S_3$ then the distance along C is 2 only if $x = c_{2r-2}$ and $y = c_{2r}$ in which case $c_{2r-2}-c_{2r-3}-\ldots c_1-c_{2r-1}-c_{2r}$ is an odd path of length $2r-1$, or if $x = c_2$ and $y = c_{2m-1}$ in which case $c_2-c_3-\ldots c_{2r-1}-c_1-c_{2m-1}$ is the desired path.

By lemma 2 S contains a pair of vertices x,y with at least $3k-2$ common neighbours. By the preceeding paragraph x and y are joined by a path $P':x-x_1-x_2-\ldots x_{2j-3}-y$ of length $2j-1$ for some $j \leq m-1$. If $j < k-1$, then there exists a set of $k+1$ vertices which are common neighbours to x and y, but which do not belong to P'; such a set is obviously a $(1,k)$-starter. If $k-1 \leq j \leq m-1$ then every neighbour z common to x and y belongs to P', since else G contains an odd cycle of length at least $2k-1$, and at most $2m-3$, contradicting the choice of C as a shortest admissible cycle. (The cycle in question would be $z-x-x_1-\ldots x_{2j-3}-y-z$.) This means that $N(x) \cap N(y) \subset V(P')$ and hence that $|N(x) \cap N(y) \cap V(C)| > 3k-2$. This is easily seen to give a contradiction, since

(1) $\qquad |N(x) \cap V(C)| < k+2$ for every vertex x in C.

Indeed, if $x = c_1$, then x can only be joined to the vertices $c_2, c_3, c_5, c_7, \ldots c_{2p-1}$, and $c_{2m-1}, c_{2m-2}, c_{2m-4}, \ldots, c_{2m-2p}$ for $2p-1 < \frac{k}{2}$. This contradiction proves case 3.

<u>Case 4</u>. $2k-1 < 2m-1$ and G contains an odd cycle C' which intersects C in a path of length $\nu(C')-2$, and C' is shorter than C. In this case C' must have length at most $\frac{k}{2}$, since else C' is a shorter admissible cycle than C.

Let $c_1-c_2-\ldots-c_{2r-2}$ be the path in common for C and C', and let v be the vertex in $V(C')-V(C)$. Put $S = \{c_i : i = 1,2,\ldots 2r\} \cup \{c_i : i = 2r+4p-1, p = 1,2,\ldots [\frac{2m-1-2r}{4}]-1\} \cup \{c_i : i = 2r+4p, p = 1,2,\ldots [\frac{2m-1-2r}{4}]-1\} \cup \{v\}$. Then S is a set of at least $k+1$ vertices, any pair of which are joined by an odd path of length at most $2m-5$, using only vertices in $\{v\} \cup V(C)$. By lemma 2 S contains a pair of vertices x,y with at least $3k-2$ common neighbours. As in the last part of case 3, $N(x) \cap N(y) \subset V(C) \cup \{v\}$. Consequently $|N(x) \cap V(C)| > 3k-3$ and $|N(y) \cap V(C)| > 3k-3$.

Since one of x,y belongs to C we again reach a contradiction to (1). This proves case 4.

Case 5. $2k-1 < 2m-1$, C is chord-free, and no vertex in $G - V(C)$ is joined to C by more than two edges.

This is clearly the last case to consider. The arguments in the proof of Theorem 1 give immediately that

$$\sum_{x \in V(C)} d(x) \leq 2n, \text{ and hence that } \delta(G) \leq [\frac{2n}{2m-1}] \leq [\frac{2n}{2k+1}],$$

contrary to assumption. This finishes the proof of lemma 4.

Theorem 2 is an immediate consequence of lemmas 3 and 4.

__Theorem 3.__ Every triangle-free graph with $\delta > [\frac{3n}{8}]$ is homomorphic with a 5-cycle.

Proof of Theorem 3. First a lemma. Let H denote the graph obtained from an 8-cycle $x_1 - x_2 - \ldots - x_8 - x_1$ by adding the edges (x_1, x_5) and (x_2, x_6). Then we have

__Lemma 5.__ Every triangle-free graph G which contains H as a subgraph fulfils $\sum_{x \in V(H)} d(x) \leq 3n$. In particular $\delta(G) \leq [\frac{3n}{8}]$.

__Proof.__ Notice that H contains no independent set of four vertices since the only such sets in $x_1 - x_2 - \ldots - x_8 - x_1$ are $\{x_2, x_4, x_6, x_8\}$ and $\{x_1, x_3, x_5, x_7\}$, neither of which is independent in H. Thus, for all $x \in V(G)$, $|E_G(x,H)| \leq 3$. It follows that

$$\sum_{x \in V(H)} d_G(x) = \sum_{x \in V(H)} |E_G(x,V(G))| = \sum_{y \in V(G)} |E_G(H,y)| \leq 3n.$$

This proves the lemma.

Let G be the given triangle-free graph in theorem 3. Since G has minimum degree larger than $\frac{2n}{7}$, Theorem 1 guarantees that G contains a 5-cycle, unless G is bipartite. In the latter case G is homomorphic with K_2 and thus with C_5, and there is nothing to prove. Hence we may assume that G contains the 5-cycle $C: v_1 - v_2 - \ldots - v_5 - v_1$. Let D be the set of vertices in G joined to vertices in C by exactly two edges. Since no vertex can be joined to C by three or more edges without forcing a non-existing triangle in G we have

(3) $\quad |D| \geq \sum_{i=1}^{5} d(v_i) - n > [\frac{7n}{8}]$ since $\delta(G) \geq [\frac{3n}{8}] + 1$.

The sets

$$D_i = N(v_{i-1}) \cap N(v_{i+1}), \quad i = 1, 2, \ldots 5$$

with indices counted modulo 5 partition D. Moreover $G[D]$ is homomorphic with C_5, since every pair of vertices x,y either belong to $D_i \cup D_{i+2}$ for some i, or else one belongs to D_i and the other to D_{i+1}, and only in the second case can x and y be joined by an edge, because $D_i \cup D_{i+2} \subset N(v_{i+1})$. Consequently we may assume that $T = V(G) - D$ is non-empty. Let x be a vertex in T. By (3), $|T| = n - |D| \leq [\frac{n}{8}] < d(x)$. It follows that x is joined to some D_i, say to D_1. We claim that

(4)   x is joined to different D_i's.

This can be seen as follows. We already know that x is joined to D_1. Consider the vertices v_3 and v_4. Since neither v_3 nor v_4 can be joined to D_1 it follows that

$$N(v_3) \cup N(v_4) \subset V(G) - D_1.$$

We also know that $N(v_3) \cap N(v_4) = \emptyset$, because of the edge (v_3, v_4), and hence

$$2[\tfrac{3n}{8}] + 2 \leq d(v_3) + d(v_4) \leq n - |D_1|.$$

Consequently $|D_1| \leq n - 2[\tfrac{3n}{8}] - 2$, and thus

$$|D_1| + |T| - 1 \leq n - 2[\tfrac{3n}{8}] + 2 + [\tfrac{n}{8}] - 1 \leq [\tfrac{3n}{8}] < d(x).$$

It follows that x must have a neighbour outside $T \cup D_1$, which proves (4).

We now consider two cases.

Case 1. x is joined to D_1 and D_2.

Then, let $x_1 \in D_1 \cap N(x)$ and $x_2 \in D_2 \cap N(x)$. The graph with vertices $\{x, x_1, x_2, v_1, v_2, v_3, v_4, v_5\}$ and edges $\{(v_i, v_{i+1}), i = 1, 2, \ldots, 5\} \cup \{(x, x_1), (x, x_2), (x_1, v_2), (x_2, v_1), (x_2, v_3)\}$ is a subgraph of G isomorphic to the graph H in lemma 5. We deduce therefore that $\delta(G) \leq [\tfrac{3n}{8}]$ contrary to assumption. Hence we have reached

Case 2. No vertex in T is joined to consecutive D_i's. In this case

$$T_i = \{x \in T: N(x) \cap D_{i-1} \neq \emptyset \text{ and } N(x) \cap D_{i+1} \neq \emptyset\}$$

is a partition of T, since every vertex in T is joined to two D_i's by (4).

We claim that

(5) $T_i \cup T_{i+2}$ is an independent set of vertices in G for $i = 1,2,\ldots 5$. To see this, fix i and assume that (x,y) is an edge between vertices in $T_i \cup T_{i+2}$. Let $x_{i+1} \in N(x) \cap D_{i+1}$ and $y_{i+1} \in N(y) \cap D_{i+1}$. Then the graph with vertices $(V(C) - v_{i+1}) \cup \{x,y,x_{i+1},y_{i+1}\}$ and edges $(E(C) - \{(v_i,v_{i+1}), (v_{i+1},v_{i+2})\}) \cup \{(v_i,y_{i+1}),(y_{i+1},v_{i+2}),(v_i,x_{i+1}),x_{i+1},v_{i+2}), (x,x_{i+1}),(x,y),$ and $(y,y_{i+1})\}$ is a subgraph isomorphic to H in G, which again leads to a contradiction. This proves (5).

It is obvious that $D_i \cup T_i$ and $D_i \cup T_{i+2}$ are independent sets of vertices for all i, since otherwise some vertex in T is joined to consecutive D_i's and this is forbidden in case 2. We deduce that G has edges between the independent sets $D_i \cup T_i$ and $D_{i+1} \cup T_{i+1}$ only, for $i = 1,2,\ldots,5$. Hence G is homomorphic to C_5.

3. REMARKS AND CONJECTURES

1. The truth of Theorem 1 prompts the following conjecture.

Conjecture 1. Every hamiltonian non-bipartite graph with minimum degree more than $\frac{2n}{5}$ is pancyclic (i.e. contains cycles of all lenghts k, $3 \leq k \leq n$).

It has been shown by Bondy [10] that every hamiltonian non-bipartite graph with at least $[\frac{n^2}{4}]$ edges is pancyclic. This was sharpened in [6] to the essentially best possible statement of its kind, namely: every hamiltonian non-bipartite graph with more than $[\frac{(n-1)^2}{4}] + 1$ edges is pancyclic.

Note however, that while conjecture 1 would be best possible for $k = 3$, there seems to be no such tight fit for $k = n-1$. In fact I know of no example of a hamiltonian non-bipartite graph with minimum degree larger than $\frac{n}{4}$ which fails to contain a C_{n-1}.

2. Erdös and Simonovits considered the function $\psi(n,K_r,t)$ defined as the maximal value of $\delta(G)$, where G is a graph on n vertices which is K_r-free and has chromatic number less than t. They

conjectured in particular that $\psi(n,K_3,t) \approx \frac{n}{3}$ for $t \geq 4$.
This conjecture must be modified slightly, since $\psi(n,K_3,4) \geq \frac{10n}{29}$ because of the graph in figure 1. It is natural to conjecture therefore that $\psi(n,K_3,4) = \frac{10n}{29}$, but the following is more likely to be true.

<u>Conjecture 2</u>. Every triangle-free graph with $\delta > \frac{10n}{29}$ is homomorphic with a Möbius ladder on 8 vertices, i.e. an 8-cycle together with the chords joining vertices of distance 4 on the cycle.

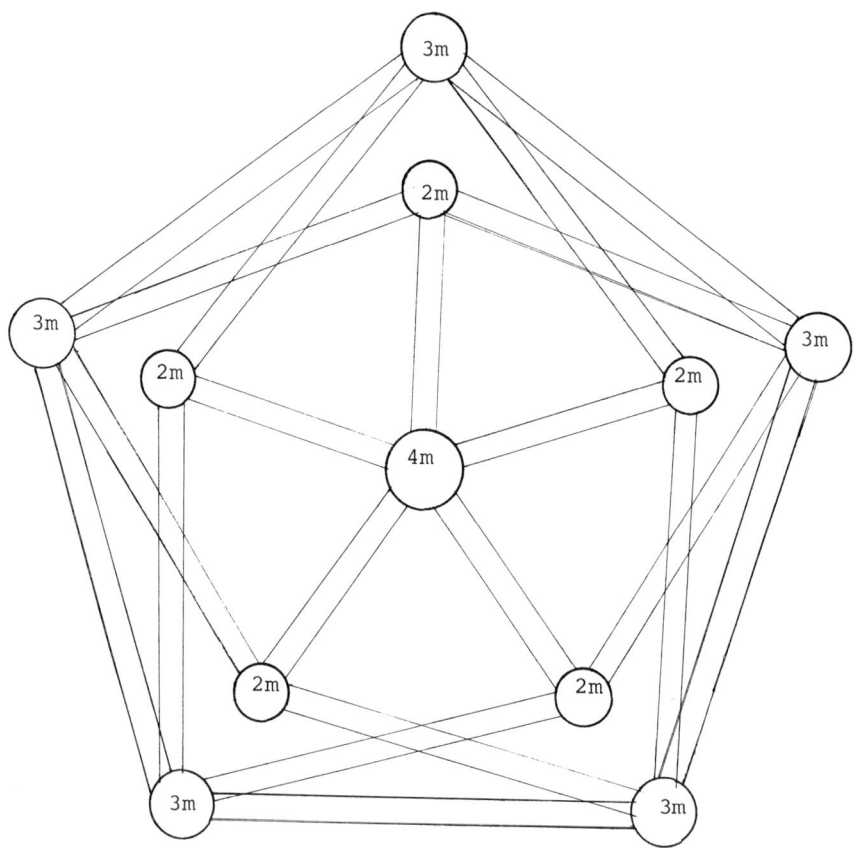

Figure 1: A 4-chromatic triangle-free graph on on 29m vertices each of degree 10m.

REFERENCES

1. B. Andrásfai, P. Erdős and V.T. Sós, On the connection between chromatic number, maximal clique and minimal degree of a graph, Discrete Mathematics 8 (1974) 205-218.

2. J.A. Bondy and U.S.R. Murty, Graph theory with applications, MacMillan, London, 1976.

3. B. Bollobás, Extremal graph theory, Academic Press, London 1978.

4. P. Erdős and M. Simonovits, On a valence problem in extremal graph theory, Discrete Mathematics 5 (1973) 323-334.

5. P. Erdős, On a theorem of Rademacher-Turán, Illinois Journal of Mathematics 6 (1962) 122-126.

6. R.J. Faudree, R. Häggkvist and R.H. Shelp, Pancyclic graphs-connected Ramsey number, submitted.

7. W. Mantel, Problem 28, solution by H. Grouwentak, W. Mantel, j.Teixeira de Mattes, F. Schuh and W.A. Wythoff, Wiskundige Opgaven 10 (1907) 60-61.

8. P. Turán, On an extremal problem in graph theory, Math. Fiz. Lapok 48 (1941) 436-452. (in Hungarian with German abstract.)

9. H.J. Voss and C. Zuluaga, Maximale gerade und ungerade Kreise in Graphen I, Wissenschaftliche Zeitschrift der Technisce Hochschule Ilmenau, Heft 4 (1977) 57-70.

10. J.A. Bondy, Pancyclic graphs I, journal of Combinatorial Theory B 11 (1971) 80-84.

SIMPLICIAL DECOMPOSITIONS:
SOME NEW ASPECTS AND APPLICATIONS

R. HALIN

Mathematisches Seminar, University of Hamburg
West Germany

1. INTRODUCTION

Decompositions and unique representation theorems arising from them play a great rôle in algebra. It is natural to ask for similar results in graph theory. There is for instance the decomposition induced by the cartesian product. Another kind of decompositions having some formal similarities with algebraic factorizations is the simplicial decomposition (abbreviated: s.d.) which is the topic of this paper. S.d.'s were the first time extensely used by K. Wagner in 1937 in his studies related to the Four-Colour-Problem. In the first place I am going to present some new results, aspects and applications which are not yet in my paper [6]; however a few basic results will be repeated here because I cannot assume familiarity with the previous paper.

2. NOTATION

Graphs are allowed to be infinite; $|G|$ denotes the cardinality of $V(G)$. $H \subseteq G$ means that H is a subgraph of G; $H \subset G$ indicates that H is a proper subgraph of G, and $H \trianglelefteq G$ that G contains H as an induced subgraph. $G \succ_u H$ expresses that G contains a subdivision of G. $x \| y$ indicates that vertices x and y are not adjacent. A complete graph will be called a simplex; we reserve the letter S for simplices $S(\alpha)$ denoting the simplex of order α. $a.T.b$ (G) means that vertices a,b are separated in G by $T \subseteq G$. $\mu_G(a,b)$ is the Menger number of vertices a,b in G, i.e. the maximum number of internally disjoint a,b-paths in G. For a cardinal $\mathfrak{m}$ its immediate successor is denoted $\mathfrak{m}^+$.

3. THE NOTION OF SIMPLICIAL DECOMPOSITION

Let G be a graph and let G_λ $(\lambda < \sigma)$ be a family of induced subgraphs of G, where σ is an ordinal > 0. We say that this family of G_λ forms a *simplicial decomposition* (s.d.) of G, if the following conditions are fulfilled:

1. $G = \bigcup_{\lambda < \sigma} G_\lambda$;
2. For all τ, $0 < \tau < \sigma$, $(\bigcup_{\lambda < \tau} G_\lambda) \cap G_\tau = S_\tau$ is a simplex;
3. Each S_τ is $\subset \bigcup_{\lambda < \tau} G_\lambda$ and $\subset G_\tau$.

Thus G is built up from the G_λ's by an infinite process (if σ is infinite) in each step of which G_λ is pasted to the part constructed before along a simplex, like an infinite cactus is composed of its branches.

It is clear from the construction that every S_λ separates G, and that for instance the chromatic number of G equals the least upper bound of the $\chi(G_\lambda)$, if σ is finite or all the S_λ are finite. (This may be wrong in the case of infinite σ and G_λ's.) The S_λ are also called the simplices of *detachment* of the s.d. in question.

4. SIMPLICIAL SUMMANDS

The subgraphs which may occur in some s.d. of a graph may be considered as something like the direct summands of a group. We call them the *simplicial summands* of the graph in question. They can be characterized in several ways.

Let us call $H \subseteq G$ *separation-invariant* in G, if

$$a.T.b\ (H) \implies a.T.b\ (G).$$

Furthermore, $H \subseteq G$ will be called *convex* in G, if for any $a,b \in V(H)$ and any chordless a,b-path $P \subseteq G$ (i.e. $P \trianglelefteq G$) also $P \subseteq H$ holds. Then we can state:

(1) Let $H \subseteq G$, H not a simplex. The following statements are equivalent:

   a) H is a simplicial summand of G;
   b) H is separation-invariant in G,
   c) H is convex in G.

Further it is easy to show

(2) The intersection of any family of convex subgraphs of G is again a convex subgraph of G.

This motivates the introduction of the *convex closure* $C_G(H)$ of some subgraph H of G as the intersection of all convex subgraphs which contain H. (Mind that G itself of course is convex in G.)

We can give an estimate of the order of $C_G(H)$ by the Menger numbers of non-adjacent vertices in G:

(3) Let $\mathfrak{m}$ be a regular cardinal $> \aleph_0$ and assume $\mu_G(x,y) < \mathfrak{m}$ for all $x \| y$, $x,y \in V(G)$. Then for any $H \subseteq G$ with $|H| < \mathfrak{m}$ we have

$$|C_G(H)| < \mathfrak{m} .$$

From this we can prove

(4) <u>First decomposition theorem</u>. Let $\mathfrak{m}$ be a regular cardinal $> \aleph_0$. Let G be a graph with $|G| \geq \mathfrak{m}$, $\mu_G(x,y) < \mathfrak{m}$ for all $x \| y$, and $S(\mathfrak{m}) \not\subseteq G$. Then G has a s.d. G_λ ($\lambda < \sigma$, where σ is the initial ordinal of $|G|$) in which all $|G_\lambda| < \mathfrak{m}$.

5. PRIME-GRAPH DECOMPOSITIONS

G is called *prime* (with respect to s.d.) if there is no s.d. of G with at least 2 members (or, equivalently, if G has no separating simplex). A s.d. G_λ ($\lambda < \sigma$) of G is called a *prime-graph decomposition* (pgd) if every G_λ is prime.

When does there exist a pgd? It is easy to show that every finite graph has a pgd. On the other hand I have an example of a countable graph which contains an infinite simplex and which does not possess a pgd. I proved:

(5) <u>Second decomposition theorem</u>. If G does not contain an infinite simplex then G has a pgd. This pgd may be chosen reduced, i.e.: no member is contained in another member. Each reduced pgd. contains exactly all the maximally prime subgraphs of G.

Thus, if no $S(\aleph_0)$ in G exists, the members of any reduced pgd are uniquely determined, though their order may vary. If $G \supseteq S(\aleph_0)$, even if a reduced pgd exists, its members need not be uniquely determined. Let us consider an example.

Let T_3 be the tree which is regular of degree 3. If a root r in T_3 is chosen, T_3 determines an order theoretical tree. Let G be the comparability graph of the partial order determined by r, i.e. vertices of T_3 are connected by an edge in G iff they are comparable (i.e. iff they lie on a path of T_3 starting in r). Then all infinite paths of T_3 starting in r determine maximal prime subgraphs of G (namely infinite simplices), and every maximal prime subgraph of G is obtained in this way. Thus G is countable and has $\aleph_1$ maximal prime subgraphs. G has a pgd with maximally prime members, which can be shown by exhausting step by step its vertices by maximally prime subgraphs; it does not matter in which order this is done. By reasons of cardinality no pgd of G can have more than $\aleph_0$ members. Therefore G has uncountably many pgd's which are essentially different.

6. A PARTIAL ORDER ON THE SET OF MAXIMAL PRIME SUBGRAPHS OF G.

One could ask whether the non-existence of pgd's for certain graphs is due to the fact that we stipulated, in pgd, a certain well-ordering of its members (or, more precisely, a tree order, i.e. a partial order in which for each x the elements preceeding x form a well-ordered set), and that perhaps, if the conditions on the order of the members in a s.d. are weakened, all graphs have a pgd. From this point of view the following relation in the set $\mathcal{P}$ of maximal prime subgraphs of G may be of interest.

For any distinct $P, Q \in \mathcal{P}$ let $S(P,Q)$ denote the simplex $\subseteq Q$ which is spanned by the vertices of Q which are adjacent to that component of G-Q which contains V(P-Q). (This component is uniquely determined, by the primity of P.) Now fix an $R \in \mathcal{P}$ (as a root), and put, for different $P, Q \in \mathcal{P} - \{R\}$, $P < Q$ iff neither $S(Q,P)$ contains $S(R,P)$ nor $S(R,P)$ contains $S(Q,P)$. Then P separates R and Q. Furthermore, put $R < P$ for each $P \in \mathcal{P} - \{R\}$.

(6) < is a partial order in $\mathcal{P}$.

For $Q \in \mathcal{P}$, the set of all $P \leq Q$ need not be a chain, but in this set "uncomparable" is an equivalence relation. It would be interesting to obtain further properties of this partial order.

7. TRIANGULATED GRAPHS

The two decomposition theorems were used in (6) to solve some colouring and connectivity problems, which are not repeated here. I now want to talk on triangulated graphs (or rigid circuit graphs). A graph G is called *triangulated* if each circuit of length ≥ 4 has a chord (i.e. if no circuits other than triangles are induced subgraphs of G). It is not difficult to prove

(7) A finite graph is triangulated iff it has a pgd consisting only of simplices.

This can be generalized by the 2nd decomposition theorem:

(7') A graph without an infinite simplex is triangulated iff it has a pgd consisting only of simplices.

Perfect graphs and s.d.'s are related by the following fact:

(8) Let $G_1,\ldots,G_r$ (r finite) be a s.d. of G. Then G is perfect iff each G_i is perfect (i = 1,...,r). (See Berge [2], p.349.)

This implies an early result of Hajnal and Suranyi:

(8') Every finite triangulated graph is perfect.

There are infinite, non-perfect triangulated graphs (see section 8 below). S. Wagon put the question [11], whether every triangulated graph without an infinite simplex is perfect. By using the 2nd decomposition theorem I could give a positive answer [7]:

(9) Every triangulated graph without an infinite simplex is perfect.

8. COMPARABILITY GRAPHS OF TREES.

A partial order $\leq$ in a set V is called *treelike* if for every $x \in V$

$$A(x) := \{v \in V | v \leq x\}$$

is a chain (with respect to $\leq$). $(V, \leq)$ is called an *order theoretical tree* if every $A(x)$ is well ordered.

The following statements for a treelike order $(V, \leq)$ are readily verified:

i) Each chain K of $(V, \leq)$ is contained in a maximal chain of $(V, \leq)$.

ii) If K is a maximal chain and $x \in K$, then $A(x) \subseteq K$.

iii) If K_i ($i \in I$) is a family of maximal chains and $x,y \in \bigcup_{i \in I} K_i$ are comparable, then there is an $i \in I$ with $x,y \in K_i$.

As a consequence we have

(10) If G is the comparability graph of a treelike order $\leq$, then there exists a pgd of G in which all the members are simplices and correspond to certain maximal chains with respect to $\leq$; the

order in which these simplices occur is arbitrary.

Proof. Well order $V(G)$, and then exhaust $V(G)$ by choosing, step by step, maximal chains.

We have seen that in the infinite case not all the maximal chains (i.e. the maximal simplices of G) may appear in the pgd. However, for finite graphs the converse of (10) also holds, and we get the following characterization of finite comparability graphs (besides the criterion of E.S. Wolk):

(11) Let G be a finite graph. The following statements are equivalent
a) G is the comparability graph of an order theoretical tree;
b) G is connected and has a pgd each member of which is a simplex such the order of these members may be chosen arbitrarily;
c) G is connected and has a pgd $G_1, \ldots, G_r$, each G_i being a simplex, such that the set of simplices of detachment contained in each G_i is a chain with respect to inclusion.

The members of the pgd just correspond to the maximal chains of the order theoretical tree in question.

By using the 1st decomposition theorem I could prove the following (χ denoting the chromatic number)

(12) $\chi(G) \geq \aleph_0 \Longrightarrow G \succ_u S(\mathfrak{m})$ for every $\mathfrak{m} < \chi(G)$.

The question whether always $G \succ_u S(\chi(G))$ for $\chi(G)$ infinite, can be considered as an infinite extension of Hadwiger's conjecture. Of course the conjecture is trivially wrong in the case that $\chi(G)$ is a limit cardinal. However the conjecture fails in the other cases as well, hence (12) is best possible. I have been led to this by an example of R. Laver (see [1]).

Let E be the set of all injective functions I of some countable ordinal δ into $\mathbb{N}$ such that $\mathbb{N} - I(\delta)$ is infinite (δ varies!). $(E, \subseteq)$ is an order theoretical tree (here $\subseteq$ is the inclusion for functions which are considered as sets). Let G denote the comparability graph of $(E, \subseteq)$. Then $\chi(G) = \aleph_1$, $G \not\succ_u S(\aleph_1)$ as is shown by some tricks.

Therefore we may say: Hadwiger's conjecture fails for infinite chromatic numbers.

9. INTERVAL GRAPHS

$G = (V, E)$ is called an *interval graph* if for each $v \in V$ there can be found an interval $I(v)$ on the real line such that

$$vv' \in E \iff I(v) \cap I(v') \neq \emptyset.$$

Let us call a finite s.d. $G_0, \ldots, G_\ell$ with the simplices of detachment S_λ a *consecutive* s.d., if $S_\lambda \subseteq G_{\lambda-1}$ for all $\lambda = 1, \ldots, \ell$ holds. Then we have

(13) A finite graph G is an interval graph iff G has a consecutive pgd consisting of simplices.

This result is closely related to the criterion of Fulkerson-Gross [3], which is however primarily formulated in the language of matrices.

It is not difficult to deduce the criterion of Gilmore-Hoffman (for the finite case) from (13). (The latter says [4] that G is an interval graph iff G is triangulated and the complement $\bar{G}$ is a comparability graph (of a suitable partial order).) Much more difficulties arise if one tries to derive the criterion of Lekkerkerker-Boland [9] from (13) and thus to obtain a simpler proof for the latter. It would be of great interest to get such a proof, in order to simplify and unify the theory of interval graphs.

The proofs of the results so far (with the exception of section 6) can be found in [8], Chap. X.

10. $\mathcal{T}$ -DECOMPOSITIONS

In the final sections I will discuss some generalizations of the s.d. The separating simplices of a graph G have the property that none of them is separated by another one. Let us call $T \subseteq G$ a *separator* of G if there are vertices x,y of G such that T is minimally separating x,y in G, i.e.: $x.T.y\,(G)$, but $\neg\, x.T-t.y\,(G)$ for every $t \in V(T)$. Let $\mathcal{T}$ be a system of separators of G. $\mathcal{T}$ is called *graded* if

1. No $T \in \mathcal{T}$ is separated by a $T' \in \mathcal{T}$;
2. If $T \in \mathcal{T}$ and $T' \subseteq T$ is a separator of G, then $T' \in \mathcal{T}$.

Obviously the separating simplices of G form an example of a graded system of separators of G.

Let $\mathcal{T}$ be a graded system of separators of G; let G be finite. We now define a $\mathcal{T}$ -decomposition of G quite analogously to a s.d.

of G; only instead of the simplices of attachment we allow now merely elements of $\mathcal{T}$.

Let $G_0, \ldots, G_\ell$ be induced subgraphs of G. We say that $G_0, \ldots, G_\ell$ (in this order) form a $\mathcal{T}$-*decomposition* of G, if

1. $\bigcup_{\lambda=0}^{\ell} G_\lambda = G$;
2. For all μ, $1 \le \mu \le \ell$, $(\bigcup_{\lambda=0}^{\mu-1} G_\lambda) \cap G_\mu = T_\mu \in \mathcal{T}$;
3. $\bigcup_{\lambda=0}^{\mu-1} G_\lambda$ and G_μ contain T_μ properly;
4. For every μ, $0 < \mu \le \ell$, there is a λ with $1 \le \lambda < \mu$ such that $T_\mu \subseteq G_\lambda$ holds. (If T_μ is a finite simplex, 4. follows from the other conditions!).

Furthermore, we call an induced subgraph H of G $\mathcal{T}$-*prime* if no pair of vertices of H is separated in G be a member of $\mathcal{T}$. A $\mathcal{T}$-decomposition of G is a $\mathcal{T}$-*prime graph decomposition* if each of its members is $\mathcal{T}$-prime. (Remark: It is not even required that each member of this decomposition be connected.) Then we can prove the following theorem:

(14) Let G be a finite graph, $\mathcal{T}$ a graded system of separators of G. Then there exists a reduced $\mathcal{T}$-prime graph decomposition, whose members must be exactly all maximal $\mathcal{T}$-prime subgraphs of G.

This result can be extended to infinite graphs to obtain a result similar to the 2nd decomposition theorem if each $T \in \mathcal{T}$ is required to be finite and that there is no properly ascending infinite chain $T_1 \subset T_2 \subset T_3 \subset \ldots$ of elements of $\mathcal{T}$.

11. ARTICULATORS

A separator T of G is called an *articulator* if for any non-adjacent vertices x,y of T there holds $\mu_G(x,y) > |T|$. It is not difficult to prove [5]:

(15) The set $\mathcal{A}$ of articulators of G forms a graded system of separators of G.

Obviously $\mathcal{A}$ contains all separating simplices of G; the latter form a graded subsystem of $\mathcal{A}$. It is also clear that, if we only take the elements of n or less elements in a graded system of separators, we again get a graded system. Hence $\mathcal{A}_n$, the set of articulators with at most n vertices, is again graded.

It is not difficult to prove that a 2-connected graph which is not 3-connected is either a circuit or has an articulator with 2 elements (see Tutte [10], theorem 11.34). Furthermore, one can prove that if G is n-connected and e is an edge such that G-e is (n-1)-separated, then G-e has at least one (n-1)-articulator. From this in [5] another successive construction of the finite 3-connected graphs is derived.

12. SEPARATING EDGE-SETS

The main notions of sections 10, 11 can be carried over also to separating edge-sets. Let $\mathcal{T}$ be a system of edge-sets of G. We call $\mathcal{T}$ a *graded system* of separating edge-sets of G, if
1. each $T \in \mathcal{T}$ is a cut of G;
2. no $T \in \mathcal{T}$ is separated by a $T' \in \mathcal{T}$;
3. if $T \in \mathcal{T}$ and $T' \subseteq T$ is a cut, then also $T' \in \mathcal{T}$.

Then we can prove a decomposition theorem analogous to (14); here an induced subgraph of G is called $\mathcal{T}$-prime if no pair of its edges can be separated in G by a member of $\mathcal{T}$. We can reduce the proof to (14) by considering the graph $\dot{G}$ which arises from G by inserting a vertex v_e on each edge e and then show that the set of all $\dot{T} := \{v_e | e \in T\}$ for $T \in \mathcal{T}$ forms a graded system of separating vertex sets in $\dot{G}$.

A cut $T \subseteq E(G)$ is called an *edge-articulator* of G, if, for any $T' \subseteq E(G)$ which separates two edges of T (in G), necessarily $|T'| > |T|$ holds. Then again we find that the set of edge-articulators forms a graded system.

I do not know whether these decompositions can be carried over to matroids or hypergraphs.

REFERENCES

[1] J.E. Baumgartner, Results and independence proofs in combinatorial set theory, Thesis, University of California 1970.

[2] C. Berge, Graphes et hypergraphes, Paris, Dunod 1973.

[3] D.R. Fulkerson and O.A. Gross, Incidence matrices and interval graphs, Pacific J. Math. 15 (1965), 835-855.

[4] P.C. Gilmore and A.J. Hoffman, A characterization of comparability graphs and interval graphs, Canad. J. Math. 16 (1964), 539-548.

[5] R. Halin, Über eine Klasse von Zerlegungen endlicher Graphen, Elektron. Informationsverarbeit. Kybernetik 6 (1970), 15-28.

[6] R. Halin, Simplicial decompositions of infinite graphs. Advances in graph theory, Annals of Discrete Mathematics 3 (1978), B. Bollobás ed., p.93-109.

[7] R. Halin, A note on infinite triangulated graphs, Discrete Math. 31 (1980), 325-326.

[8] R. Halin, Graphentheorie II. Wissenschaftl. Buchgesellschaft Darmstadt, to appear.

[9] C. Lekkerkerker and J. Boland, Representation of a finite graph by a set of intervals on the real line, Fundam. Math. 51 (1962), 45-64.

[10] W.T. Tutte, Connectivity in graphs, Univ. of Toronto Press, Toronto-London 1966.

[11] S. Wagon, Infinite triangulated graphs, Discrete Math. 22 (1978), 183-189.

ACHIEVEMENT AND AVOIDANCE GAMES FOR GRAPHS

FRANK HARARY[1]

University of Michigan
Ann Arbor, U.S.A.

Consider the following games played on the complete graph K_p by two players called Alpha and Beta. First, Alpha colors one of the edges of K_p green, then Beta colors a different edge red, and so forth. A graph F with no isolated points is given. The first player who can complete the formation of F in his color is the winner. The minimum p for which Alpha can win regardless of the moves made by Beta is the achievement number of F. In the corresponding misère game, the player who first forms F in his color is the loser and the minimum p for which Beta can force Alpha to lose is the avoidance number of F. Several variations on these games are proposed, and partial results presented.

0. PRELIMINARIES.

The *classical ramsey number* $r(K_m, K_n)$ is the smallest positive integer p such that every 2-coloring of $E(K_p)$, in green and red without loss of generality, must contain a green K_m or a red K_n. Its existence is assured by the celebrated theorem of Ramsey [27]. This number was previously denoted by $r(m,n)$, as for example in [2,8,10]. In November 1962, while listening to a lecture [6] by Paul Erdös on the numbers $r(m,n)$, I formulated the "generalized ramsey number" $r(F,H)$ for any two graphs F and H, not necessarily complete, with no isolated points. When $H = F$, we obtain the (diagonal) *ramsey number* $r(F) = r(F,F)$. Chvátal and I [5] whimsically defined a "small graph" as one with at most four points and no isolates and we noted their ramsey numbers.

---
[1] Overseas Fellow 1980-81, Churchill College, University of Cambridge.

Achievement and avoidance games and numbers

We only consider graphs F with no isolates. The *game of achieving* F on K_p is written $(F,K_p,+)$ and is played as follows. Beginning with p isolated vertices, the first player (Alpha or A) draws a green line segment joining two of them, the second (Beta or B) a red line joining another pair of vertices, etc. The player (if any) who first constructs a copy of F in his color is the winner.

In the *avoidance game* $(F,K_p,-)$, the first player (if any) who forms F in his color loses. These games were first mentioned in [11].

The *achievement number* $a(F)$ is the minimum p for which Alpha can win the game $(F,K_p,+)$ regardless of what moves are made by Beta. The *avoidance number* $\bar{a}(F)$ is the smallest p for which Beta can play in a way that forces the formation of a green F regardless of the moves made by Alpha.

Let $a = a(F)$, $\bar{a} = \bar{a}(F)$ and $r = r(F)$. It is obvious that $a \leq r$ and $\bar{a} \leq r$, as neither of the games $(F,K_{r(F)},+)$ or $(F,K_{r(F)},-)$ can end in a draw.

<u>Problem 0.</u> Determine $a(F)$ and $\bar{a}(F)$ for various families of graphs. This appears hard even for trees.

1. AVOIDING F ON r(F) POINTS.

The first graph avoidance game was called SIM by SIMMONS in a note [29] which asked for the outcome and best strategy for the game $(K_3,K_6,-)$. Later both Simmons and others proved by both personal exhaustion and with computer assistance the following definitive result.

<u>Theorem 1.</u> In the triangle avoidance game $(K_3,K_6,-)$, if both players play rationally in a way that delays a loss as long as possible, then the first player Alpha will lose on the last possible (fifteenth) edge of K_6, at which time he will form *two* green triangles.

<u>Conjecture 1.</u> Theorem 1 remains true when "triangle" and K_3 are replaced by "quadrilateral" and C_4.

Clearly this conjecture can be readily settled with computer assistance.

2. ACHIEVEMENT AND AVOIDANCE NUMBERS FOR SMALL GRAPHS.

We now give the values of a and $\bar{a}$ for those small graphs for which they have been determined by exhaustion, as mentioned in [13].

F	K_2	P_3	$2K_2$	P_4	$K_{1,3}$	K_3	C_4	$K_{1,3}+e$	K_4-e	K_4
a	2	5	5	5	5	5	6	5	?	?
$\bar{a}$	2	3	5	5	5	5	6	5	?	?
r	2	3	5	5	6	6	6	7	10	18

Shader and Cook [27] gave a strategy for Beta to force Alpha to lose the avoidance game $(K_{1,3}+e, K_6,-)$. They thus showed that $\bar{a}(F) < r(F)$ when $F = K_{1,3}+e$.

<u>UP2</u> (Unsolved Problem 2) Complete Table 2 by determining a and $\bar{a}$ for both K_4-e and K_4.

3. ECONOMICAL GRAPHS.

When $p = r(F)$, the smallest possible number of green moves required to form F in the achievement game $(F,K_p,+)$ is called the *move number*, $m(F)$. The *size* of F is the number of lines, $q(F)$. We call graph F *economical* if $m(F) = q(F)$. The economical small graphs are K_2, P_3, $2K_2$, P_4, $K_{1,3}$ and $K_{1,3}+e$. It appears that all paths P_n are economical.

Call F *ultimately economical* if for some p, Alpha can make a green F in only $q(F)$ moves. It is easy to see that all trees and hence all forests are ultimately economical, as noted by Curtis Clark (unpublished).

<u>UP3</u> Which graphs F are economical? (As this depends on the previous determination of $a(F)$, it is more reasonable to ask which trees and which forests are economical.) Which graphs are ultimately economical?

4. INEQUALITIES.

<u>Conjecture 4</u>. It seems intituitively obvious that for all F we will have $a \leq \bar{a}$. But there is no proof or disproof as yet.

If this is true as we believe, then we would be able to partition all graphs F into the following four classes.

I. $a = \bar{a} = r$, as for C_4 with $r = 6$.

II. $a = \bar{a} < r$, as for $K_{1,3}$ with $a = 5$, $r = 6$.

III. $a < \bar{a} = r$, which I think is an empty class.

IV. $a < \bar{a} < r$, which is probably satisfied by K_4 and perhaps also by K_4-e.

5. FIRST PLAYER ACHIEVEMENT.

Béla Bollobás suggested a simpler achievement game in which only Alpha aims to form a green F, and the goal of Beta is to try to stop him.

<u>Question 5</u>. For which graphs F does the achievement number under these new rules equals the previous number a(F)?

Bollobás and Papaioannou [3] have determined these achievement numbers for the paths P_n (the answer being n unless n = 4), the "bars" nK_2, and other families of graphs. Perhaps surprisingly, they proved that the hamiltonian cycle C_n can be achieved on K_n when $n \geq 8$. (See also Papaioannou [26] for a weaker result in this direction.)

The corresponding avoidance games, where only Alpha tries to avoid forming a green F, have not yet been studied.

6. GAMES ON NONCOMPLETE GRAPHS.

The *size ramsey number* $\hat{r}(F)$ has been defined both by Erdös, Faudree, Rousseau and Schelp [7] and independently by Miller and me [21] as the minimum size q(G) of a graph G such that $G \rightarrow F$, meaning that every 2-coloring of E(G) must have a monochromatic F. This suggests achievement and avoidance games played on graphs G which are not complete, which are denoted similarly by (F,G,+) and (F,G,-).

In view of the fact that F does not necessarily require r(F) points to be achieved on a complete graph, it is also meaningful to consider playing (F,H,+) for graphs $H \not\rightarrow F$. Defining the *size achievement and avoidance numbers* as expected, several questions suggest themselves.

<u>Open Question 6</u>. Determine the size achievement and avoidance numbers of small graphs, trees, and several families of graphs.

7. INDUCED SUBGRAPHS.

We write $F \subset G$ when F is a subgraph of G and $F < G$ when F is an induced subgraph. As usual, a *factorization* of G into two spanning subgraphs $G_1 \cup G_2$ is determined by a partition $E = E_1 \cup E_2$ of $E = E(G)$. In these terms, $G \to F$ if for every factorization $G = G_1 \cup G_2$, we have $F \subset G_1$ or $F \subset G_2$.

We now write $G \twoheadrightarrow F$ to mean that for every factorization $G = G_1 \cup G_2$, we have $F < G_1$ or $F < G_2$. A special case of a basic result of Nesetril and Rödl [25] proves that for each F, there exists G such that $G \twoheadrightarrow F$; see also Graham, Rothschild and Spencer [8].

This fact suggests several questions involving achievement and avoidance games. Call G_1 smaller than G_2 if $p_1 < p_2$. Define the induced ramsey *number* ri(F) as the minimum p such that for some G with p points, $G \twoheadrightarrow F$. Similarly the *induced achievement number* ai(F) is the minimum order of a graph H such that Alpha can construct a green induced subgraph F of H (by alternately coloring edges of H green and red as for K_p) regardless of the moves made by Beta. The corresponding number $\overline{ai}(F)$ is defined as expected.

Open Question 7. For each graph F, determine the induced numbers ri(F), ai(F) and $\overline{ai}(F)$. (We have already found most of these numbers for small graphs and stars, and are studying trees and various families of graphs.)

8. TREES.

A well known adage in graph theory says that when a problem is new and does not reveal its secret readily, it should first be studied for trees where it will generally be easier to handle. A companion motto urges that each question for graphs also be specialized to bipartite graphs and generalized to directed graphs. Thus this section discusses trees and the next two deal with bigraphs and digraphs.

Conjecture 8. The minimum value of a(T) among all trees T of order n is realized when $T = P_n$, the path. The maximum of a(T) is attained for T the star $K(1, n-1)$.

Related questions ask for a list of all the n-point trees for which a(T) is minimum or maximum, for the set of all values of a(T), and corresponding results for the avoidance number. The conjecture when a(T) is replaced by r(T) seems to be true.

We remark that both trees of order 4 have a(T) = 5 and that the three trees of order 5 have a(T) = 5, 5 and 7.

9. BIGRAPHS.

When both F and G are specialized to bipartite graphs in Section 6 on noncomplete graphs, we get achievement and avoidance games on bigraphs. One such game was analyzed in [14], namely $(K_{1,4}, K_{5,5}, +)$, where it was shown that the move number is 8. It is also easy to verify that the move number of $(C_4, K_{4,4}, +)$ is 7.

The associated ramsey theory for bigraphs, proposed in [12], was developed by Harborth, Mengersen and me [19] where we defined the *bipartite ramsey set* of a bigraph F as the collection of all ordered pairs (m,n) with m ≤ n such that $K_{m,n} \to F$ but $K_{m-1,n}, K_{m,n-1} \not\to F$. Achievement games for bigraphs are being investigated by Exoo and myself. Bipartite achievement (and analogously avoidance) numbers can either be defined as the minimum n such that $K_{n,n} \xrightarrow{+} F$ (where the notation as expected means that Alpha can make a green F playing on $K_{n,n}$) or as the minimum p = m + n such that $K_{m,n} \xrightarrow{+} F$.

<u>Open Question 9</u>. Determine bipartite achievement and avoidance numbers for trees (especially stars and paths), for even cycles, and for other families of bipartite graphs.

10. DIGRAPHS.

The digraph $\underline{D}G$ of a graph G is defined to have the same point set V(G) with each edge of G replaced by a symmetric pair of arcs. For a given digraph D, again with no isolates, the *ramsey number* r(D), if any, is the minimum p such that $\underline{D}K_p \to D$, where the arrow now refers to the coloring of arcs. It was independently and simultaneously discovered by Bermond [1] and by Hell and me [20] that r(D) exists if and only if D is acyclic.

It is easy to see that for the cyclic triple C_3, Alpha can win the achievement game $(C_3, \underline{D}K_5, +)$ even though C_3 is not acyclic and that the move number is 5.

Obviously no digraph containing a symmetric pair of arcs can be achieved. However, Andrew Thomason has proved that any asymmetric digraph (oriented graph) D can be achieved on $\underline{D}K_p$ for some p.

The corresponding statement for avoidance is still open.

Question 10. What are the achievement and move numbers for directed paths and directed cycles?

11. MULTIPLICITY GAMES.

The *ramsey multiplicity* of F, written $R(F)$, is the minimum possible number of monochromatic copies of F in any 2-coloring of K_n where $n = r(F)$. This concept was introduced in [23] where $R(F)$ was determined exactly for all small F except K_4. A survey of results subsequently derived about $R(F)$ is given by Burr and Rosta [4].

A *multiplicity game* for F is played like an achievement (or an avoidance game) except that the winner (or loser) is not the first player to complete a monochromatic F, but the player who completes in his color the most (least) copies of F.

Open Question 11. Let F be a graph with $n = r(F)$. What are the outcomes of the F-multiplicity achievement and avoidance games played on K_n?

Of course similar questions can be asked for digraphs, bipartite graphs, induced subgraphs, and so forth.

12. CONCLUDING REMARKS.

Many other achievement and avoidance games suggest themselves naturally. These include games designed from theorems [15] not only about graphs as in the present communication, but also concerning number theory, group theory, Latin squares, combinatorial designs and finite geometries.

All the games about graphs presented above involve coloring the edges using two colors. This procedure can be modified in two ways. First, the vertices can be colored instead, as pointed out by J. Nešetřil and V. Rödl (verbal communication). Second, the number of colors may be different from two. When only one color is used, we obtain games on n points, with objective graph F having order at most n, in which both players A and B use white chalk and the first player to complete F either wins or loses. When the identification of subgraph F is an NP-hard problem, e.g., if F is a hamiltonian cycle C_n, then these have been called "hard games" by M. Grötschel (verbal communication).

It is well known [9] that the 3-color ramsey number of a triangle, written $r(K_3,3)$, is 17. Nevertheless it has been found empirically that when three players A, B and Γ play triangle achievement (using blue, green and red chalk without loss of generality) on only seven points, the result does not terminate in a drawn game.

Other graphical concepts also provide interesting material for games, such as convexity [22,17], connectivity, colorability, and so forth.

Several combinatorial settings which are not explicitly graph theoretic are eminently appropriate for the design and play of achievement and avoidance games. Thus there are games involving chess pieces [18,24], games with square-celled animals [16], and games on triangular animals and polyhexes being explored with Heiko Harboth.

REFERENCES.

1. J.-C. Bermond, Some ramsey numbers for directed graphs. *Discrete Math.* 9 (1974) 313-321.

2. B. Bollobás, *Extremal Graph Theory*. Academic Press, London (1978).

3. B. Bollobás and I. Papaionnou, Achievement numbers for graphs, to appear.

4. S.A. Burr and V. Rosta, On the ramsey multiplicities of graphs -- Problems and recent results. *J.Graph Theory* 4 (1981) 347-361.

5. V. Chvátal and F. Harary, Generalized ramsey theory for graphs II: Small diagonal numbers. *Proc. Amer. Math. Soc.* 32 (1972) 389-394.

6. P. Erdős, Applications of probabilistic methods to graph theory. A *Seminar on Graph Theory* (F. Harary, ed.), Holt, New York (1967) 60-64.

7. P. Erdős, R.J. Faudree, C.C. Rousseau, R.H. Schelp, The size ramsey number. *Periodica Math. Hungar.* 9 (1978) 145-161.

8. R.L. Graham, B. Rothschild, J. Spencer, *Ramsey Theory*. Wiley, New York (1981).

9. R.E. Greenwood and A.M. Gleason, Combinatorial relations and chromatic graphs. *Canad. J. Math.* 7 (1955) 1-7.

10. F. Harary, *Graph Theory*. Addison-Wesley, Reading (1969).

11. F. Harary, Recent results on generalized ramsey theory for graphs. *Graph Theory and Applications*. Springer Lecture Notes 303 (1972) 125-138.

12. F. Harary, The foremost open problems in generalized ramsey theory. *Proc. Fifth British Combinatorial Conference*. Utilitas, Winnipeg (1976) 269-282.

13. F. Harary, Generalized ramsey theory for graphs I to XIII. *The Theory and Application of Graphs* (G. Chartrand et al. eds.) Wiley, New York (1981), to appear.

14. F. Harary, An achievement game on a toroidal board. *Proceedings - First Polish Symposium on Graph theory* (M. Sysło, ed.) Springer Lecture Notes, to appear.

15. F. Harary, Achievement and avoidance games designed from theorems. *Rendiconti Sem. Mat. Fis. Milano* (1982), to appear.

16. F. Harary, Achieving the Skinny animal. *Eureka* (1982), to appear.

17. F. Harary, Convexity in graphs: Achievement and avoidance games. *Annals Discrete Math.* (1982), to appear.

18. F. Harary, Achievement and avoidance games involving chess pieces. *J. Recreational Math.*, to appear.

19. F. Harary, H. Harborth, I. Mengersen, Generalized ramsey theory for graphs XII: Bipartite ramsey sets. *Glasgow Math. J.* 22 (1981) 31-41.

20. F. Harary and P. Hell, Generalized ramsey theory for graphs V: The ramsey number of a digraph. *Bull. London Math. Soc.* 6 (1974) 175-182; 7 (1975) 87-88.

21. F. Harary and Z. Miller, Generalized ramsey theory VIII: The size ramsey number of small graphs. *Studies in Pure Mathematics in Memory of Paul Turán.* To appear.

22. F. Harary and J. Nieminen, Convexity in graphs. *J. Differential Geometry* (1981), to appear.

23. F. Harary and G. Prins, Generalized ramsey theory for graphs IV: the ramsey multiplicity of a graph. *Networks* 4 (1974) 163-173.

24. F. Harary and A. Thomason, Covering with independent knights. *J. Recreational Math.*, to appear.

25. J. Nešetřil and V. Rödl, The ramsey property for graphs with forbidden complete subgraphs. *J. Combin. Theory* B20 (1976) 243-249.

26. A. Papaioannou, A Hamiltonian game. *Graph Theory* (B. Bollobás, ed.), Annals of Discrete Math. (1982), to appear.

27. F.P. Ramsey, On a problem of formal logic. *Proc. London Math. Soc.* 30 (1930) 264-286.

28. L.E. Shader and M. Cook, A winning strategy for the second player in the game Tri-tip. Proc. Tenth S.E. *Conference on Computing, Combinatorics and Graph Theory.* Utilitas, Winnipeg (1980).

29. G.J. Simmons, The game of SIM. *J. Recreational Math.* 2 (1969) 66.

EMBEDDING INCOMPLETE LATIN RECTANGLES

A.J.W. HILTON

University of Reading
England

Cruse gives necessary and sufficient conditions for embedding saturated $r \times s$ latin rectangles on n symbols inside latin squares on t symbols, generalizing a theorem of Ryser. We give another proof of this result as well as a number of analogues and generalizations.

1. INTRODUCTION

An $r \times s$ *partial latin rectangle* R *on symbols* $\sigma_1, \ldots, \sigma_n$ is an $r \times s$ matrix in which some of the cells contain a symbol chosen from $\{\sigma_1, \ldots, \sigma_n\}$, whilst other cells may be empty, and which satisfies the conditions: each symbol occurs at most once in each row and at most once in each column. An $r \times s$ partial latin rectangle on $\sigma_1, \ldots, \sigma_n$ is *saturated* if none of $\sigma_1, \ldots, \sigma_n$ can be placed in any of the empty cells without violating the above condition. If there are no empty cells then we call R an *incomplete* $r \times s$ *latin rectangle*.

Let $N_R(\sigma_i)$ denote the number of times that the symbol σ_i occurs in R. Ryser [12] proved the following theorem about completing latin rectangles.

Theorem 1.1 An $r \times s$ incomplete rectangle R on symbols $\sigma_1, \ldots, \sigma_t$ can be completed to a $t \times t$ latin square on the same symbols if and only if

$$N_R(\sigma_i) \geq r + s - t \qquad (1 \leq i \leq t).$$

Let D_R denote the number of empty cells in a partial $r \times s$ latin rectangle R and let D_ρ, D_c be the number of empty cells in row ρ and column c of R for $1 \leq \rho \leq r$ and $1 \leq c \leq s$. Cruse [7] extended Ryser's theorem to the following.

Theorem 1.2 A saturated $r \times s$ partial latin rectangle R on symbols $\sigma_1, \ldots, \sigma_n$ can be completed to a $t \times t$ latin square T on symbols $\sigma_1, \ldots, \sigma_t$ if and only if the following three conditions hold:

(i) $D_R \geq (r + s - t)(t - n)$,
(ii) $N_R(\sigma_i) \geq r + s - t$ $\quad (1 \leq i \leq n)$,
(iii) $D_\rho \leq t - n$ and $D_c \leq t - n$ $\quad (1 \leq \rho \leq r, 1 \leq c \leq s)$.

If $t = n$ then Theorems 1.1 and 1.2 coincide. We give a new simpler proof of Theorem 1.2.

If R is not saturated then the conditions of Theorem 1.2 are sufficient but not necessary for R to be completable to a $t \times t$ latin square T.

If $n = r = s$ then condition (i) becomes $D_R \geq (2n-t)(t-n)$, an easily sketched quadratic function of t, with leading term $-t^2$. Clearly, for some R, if t is just greater than n the inequality may be satisfied and R can be completed to T, then, for somewhat larger values of t, the inequality may not be satisfied, but then for yet larger values of t the inequality may be satisfied again.

Later in this paper we discuss symmetrical analogues of this theorem and generalizations to (p, q, x) - latin rectangles. Whilst each theorem and proof has its own individual features, there is usually a lot of similarity between its proof and our proof of Theorem 1.2, and so we generally shorten the account, where possible, by saying, for example, that our proof of (x) in Theorem X is very similar to the proof of (y) in Theorem Y.

In the analogues and generalizations it usually remains true that the conditions listed are sufficient but not necessary for the embedding if R is not saturated, and also one of the conditions is that D_R has to be at least as large as an expression which is quadratic in t with leading term $-t^2$.

In Section 2 we collect all the results on edge-colourings of graphs we need to use. In Section 3 we prove Theorem 1.2. In Section 4 we consider symmetric analogues of Theorem 1.2. In Section 5 we consider generalizations of Theorem 1.2 to (p, q, x) - latin rectangles and in Section 6 we consider symmetric (p, p, x) - latin squares. Finally in Section 7 we consider an analogue to the case where the diagonal of T is almost completely specified.

2. EDGE-COLOURING THEOREMS

An *edge-colouring* of a multigraph G is a map $f : E(G) \to \{C_1, C_2, \ldots\}$ where $\{C_1, C_2, \ldots\}$ is a set of colours. It is a *proper* edge-colouring if $f(e_1) \neq f(e_2)$ whenever e_1 and e_2 have a common vertex. The least number of colours for which G has a proper edge-colouring is denoted by $\chi'(G)$. Let $\Delta = \Delta(G)$ be the maximum degree of G and let $m = m(G)$ be the maximum multiplicity of an edge - i.e. the greatest number of edges joining any pair of vertices. For an edge-colouring of G, for each $v \in V(G)$, let $C_i(v)$ be the set of edges incident with v of colour C_i and, for $u, v \in V(G)$, $u \neq v$, let $C_i(u, v)$ be the set of edges joining u, v coloured C_i.

An edge colouring is *equitable* if

$$||C_i(v)| - |C_j(v)|| \leq 1 \qquad (\forall\ v \in V(G),\ \forall\ i, j),$$

and it is *balanced* if also

$$||C_i(u, v)| - |C_j(u, v)|| \leq 1 \qquad (\forall\ u, v \in V(G),\ \forall\ i, j).$$

de Werra ([14], [17], [18]) showed that, for any k, a bipartite graph has a balanced edge-colouring with k colours.

If k is even then it is known (see [20]) that any multigraph G has an edge-colouring with $\lceil \frac{\Delta(G)}{k} \rceil$ colours in which each vertex has at most k edges of each colour on it. (This is an easy consequence of a theorem of Petersen [11]). The well known theorem of Vizing [13] states that $\Delta(G) \leq \chi'(G) \leq \Delta(G) + m(G)$. By identifying colours it follows that if k is odd and G is a simple graph then G can be edge-coloured with k colours so that at most $\lceil \frac{\Delta(G) + 1}{k} \rceil$ edges of each colour are incident with each vertex. A graph is said to be of Class 1 if $\chi'(G) = \Delta(G)$, of Class 2 otherwise.

An edge-colouring is *equalized* if the number of edges of colour C_i differs from the number of edges of colour C_j by at most 1 ($\forall\ i, j$). Any of the types of colouring considered here can be equalized by a very simple argument (McDiarmid [10], de Werra [16]).

3. OUR PROOF OF THE PROTOTYPE THEOREM

In this section we prove Theorem 1.2, the prototype for the remaining theorems of this paper.

Proof of Theorem 1.2

1. **Necessity.**

(i) Each of the symbols $\sigma_{n+1},\ldots,\sigma_t$ must be placed sufficiently often in R so that the Ryser condition

$$N_R(\sigma_i) \geq r + s - t \qquad (n + 1 \leq i \leq t)$$

is satisfied. The number of such symbols is $(t-n)$. Consequently, the number of empty cells must be at least $(r+s-t)(t-n)$.

(ii) Since R is saturated, $\sigma_1,\ldots,\sigma_n$ must each satisfy the Ryser condition.

(iii) The vacant cells in any row or column have to be filled with the symbols $\sigma_{n+1},\ldots,\sigma_t$, so there must be at most $t-n$ of them.

2. **Sufficiency.** Suppose (i), (ii) and (iii) hold. Consider the bipartite graph G on vertices $\rho_1,\ldots,\rho_n, c_1,\ldots,c_n$ formed by joining ρ_i to c_j by an edge if and only if cell (i, j) is empty. The graph has at least $(r+s-t)(t-n)$ edges (by (i)) and has maximum degree $\leq t - n$ (by (iii)). By König's theorem, it can be properly edge-coloured with $t-n$ colours $\sigma'_{n+1},\ldots,\sigma'_t$, and by the theorem of McDiarmid and de Werra this proper edge-colouring can be equalized, and then there will be at least $r + s - n$ edges of each colour. We can then put σ_k into cell (i, j) whenever $\rho_i c_j$ is coloured σ'_k. Then the Ryser condition will be obeyed by all the symbols, so, by Ryser's theorem, the embedding can be carried out.

4. SYMMETRIC ANALOGUES OF THEOREM 1.2

In this section we make use of the following symmetric analogue of Theorem 1.2, due to Cruse [6].

Theorem 4.1 A symmetric incomplete $r \times r$ latin square R on t symbols $\sigma_1,\ldots,\sigma_t$ can be completed to a symmetric $t \times t$ latin square on the same symbols if and only if

$$N_R(\sigma_i) \geq 2r - t \qquad (1 \leq i \leq t)$$

and

$$N_R(\sigma_i) \not\equiv t \pmod{2} \quad \text{for at most} \quad t - r \text{ symbols } \sigma_i.$$

Given a symmetric $r \times r$ partial latin square R on n symbols let G be the graph formed as follows: Let the vertices be $v_1,\ldots,v_r$ and let v_i and v_j be joined by an edge if cells (i, j) and (j, i) are empty. Let G* be the graph formed from G by adjoining one further vertex v^* and the edge v^*v_i whenever cell (i, j) is empty $(1 \leq i \leq r)$. Let G' be the graph formed from G by adjoining a vertex w_i and an edge $v_i w_i$ for each i, $1 \leq i \leq r$, such that cell (i, i) is empty.

Recall that a graph G is of Class 1 if $\Delta(G) = \chi'(G)$ and is of Class 2 otherwise.

The results we obtain for the symmetric case are not so satisfactory as those of Theorem 1.2 because the necessary and sufficient conditions are in terms of G^* or G'. However fairly good numerical conditions which are sufficient for the embedding to be possible are given in the corollaries.

First we consider the case when t is odd.

<u>Theorem 4.2</u>   Let t be odd. A saturated symmetric $r \times r$ partial latin square on n symbols $\sigma_1, \ldots, \sigma_n$ can be embedded in a symmetric $t \times t$ latin square T on t symbols $\sigma_1, \ldots, \sigma_t$ if and only if

   (i) $D_R \geq (2r-t)(t-n)$
   (ii) $N_R(\sigma_i) \geq 2r - t$ $(1 \leq i \leq n)$,
   (iii)   Each symbol from $\{\sigma_1, \ldots, \sigma_n\}$ occurs at most once on the diagonal of R,
   (iv)   either $\Delta(G^*) < t - n$
           or $\Delta(G^*) = t - n$ and G^* is of Class 1.

<u>Proof</u>.

1. <u>Necessity</u>.

   (i), (ii). The proof of these is as in the proof of Theorem 1.2.
   (iii). If T is symmetric and t is odd each symbol must occur exactly once on the diagonal of T. Therefore, in R, each symbol $\sigma_1, \ldots, \sigma_n$ occurs at most once on the diagonal.
   (iv). If R is embedded in T then we can obtain a proper edge-colouring of G^* with colours $\sigma'_{n+1}, \ldots, \sigma'_t$ as follows: edge $v_i v_j (i \neq j)$ is coloured with σ'_k if symbol σ_k is placed in cell (i, j) and edge $v^* v_i$ is coloured with σ'_k if σ_k occurs in cell (i, i). The edge-colouring is proper because (i) no two cells in the same row contain the same symbol and (ii) the symbols occurring in the diagonal cells of R are all different. The number of colours used is $t - n$. Therefore, either $\Delta(G^*) = t - n$ and G^* is of Class 1 or $\Delta(G^*) < t - n$.

2. Sufficiency.

Suppose (i), (ii), (iii) and (iv) hold. G^* can be properly edge-coloured with $t - n$ colours $\sigma'_{n+1}, \ldots, \sigma'_t$. This colouring can be equalized, and then there will be at least $\frac{1}{2}(2r - t + 1)$ edges of each colour. Then, if $i \neq j$, σ'_k is placed in cells (i, j) and (j, i) if edge $v_i v_j$ is coloured with σ'_k and in cell (i, i) if edge $v^* v_i$ is coloured σ'_k. Then $N_R(\sigma_i) \geq 2r - t$ for each i, $1 \leq i \leq t$. Furthermore, no symbol will occur more than once on the diagonal of R. By Theorem 4.1, therefore, R can be completed to a symmetric latin square.

<u>Corollary 4.3</u>. Let t be odd. A symmetric $r \times r$ partial latin square on n symbols $\sigma_1, \ldots, \sigma_n$ can be embedded in a symmetric $t \times t$ latin square on t symbols if

(i) $D_R \geq (2r-t)(t-n)$,
(ii) $N_R(\sigma_i) \geq 2r - t$ $(1 \leq i \leq r)$,
(iii)   Each symbol from $\{\sigma_1, \ldots, \sigma_n\}$ occurs at most once on the diagonal of R,
(iv) $D_\rho < t - n$ $(1 \leq \rho \leq r)$.

If t is even, the result we obtain is even less pleasant. Let y be the number of symbols from $\{\sigma_1, \ldots, \sigma_n\}$ which occur an odd number of times on the diagonal of R. Let the edges of $G' \setminus G$ be called *pendant edges* of G'. In a proper edge-colouring of G' in which σ'_i is one of the colours, let $e(\sigma'_i)$ be the number of edges of G coloured σ'_i and let $p(\sigma'_i)$ be the number of pendant edges coloured σ'_i.

<u>Theorem 4.4</u>. Let t be even. A saturated symmetric $r \times r$ partial latin square R on n symbols $\sigma_1, \ldots, \sigma_n$ can be embedded in a symmetric $t \times t$ latin square T on t symbols $\sigma_1, \ldots, \sigma_t$ if and only if

(i) $D_R \geq (2r-t)(t-n)$,
(ii) $N_R(\sigma_i) \geq 2r - t$,
(iii) $y \leq t - n$,
(iv)   the edges of G' can be properly edge-coloured with $t - n$ colours, say $\sigma'_{n+1}, \ldots, \sigma'_t$, such that
       (a) $2e(\sigma'_i) + p(\sigma'_i) \geq 2r - t$ $(n+1 \leq i \leq t)$,
       (b) at most $t - n - y$ colours occur an odd number of times on the pendant edges.

Proof.

1. **Necessity.**

(i) and (ii). The proofs of these are as in the proof of Theorem 1.2.

(iii). In T each symbol occurs an even number of times on the diagonal. Therefore, since R is saturated, each symbol from $\sigma_1, \ldots, \sigma_n$ which occurs an odd number of times on the diagonal of R must occur at least once on the diagonal cells of T outside R, and so $y \leq t-n$.

(iv). If R is embedded in T then we can obtain a proper edge-colouring of G' with colours $\sigma'_{n+1}, \ldots, \sigma'_t$ as follows: edge $v_i v_j$ ($i \neq j$) is coloured with σ'_k if symbol σ_k occurs in cell (i, j) and edge $v_i w_i$ is coloured with σ'_k if σ_k occurs in cell (i, i). Each edge is coloured since R is saturated, and the edge-colouring is clearly proper. Since $N_R(\sigma_i) \geq 2r - t$ (by Theorem 4.1) it follows that

$$2e(\sigma'_i) + p(\sigma'_i) \geq 2r - t \qquad (n+1 \leq i \leq t).$$

Since the number of symbols which occur on the diagonal of R an odd number of times before and after R is filled is y and at most $t - n$ respectively (by Theorem 4.1), the number of colours occurring on an odd number of pendant edges is at most $t - n - y$.

2. **Sufficiency.**

Suppose (i), (ii), (iii) and (iv) hold. Let the empty cells of R be filled as follows: if edge $v_i v_j$ ($i \neq j$) is coloured σ'_k, then place symbol σ_k in cells (i, j) and (j, i), and if edge $v_i w_i$ is coloured σ'_k then place symbol σ_k in cell (i, i). Then R is filled symmetrically. Condition (iv(a)) implies that $N_R(\sigma_i) \geq 2r - t$ ($n+1 \leq i \leq t$) and condition (iv(b)) implies that at most $t - n$ colours occur an odd number of times on the diagonal of R. By Theorem 4.1, therefore, R can be embedded in a symmetric latin square of side t.

Corollary 4.5. Let t be even. A symmetric $r \times r$ partial latin square on n symbols $\sigma_1, \ldots, \sigma_n$ can be embedded in a symmetric $t \times t$ latin square on t symbols $\sigma_1, \ldots, \sigma_t$ if

(i) $D_R \geq (2r-t)(t-n)$,

(ii) $N_R(\sigma_i) \geq 2r - t$ $\qquad (1 \leq i \leq r)$,

(iii) $y \leq t-n$

(iv) $D_\rho \leq t - n - 2$ $\qquad (1 \leq \rho \leq r)$.

Proof. From G' form a graph G'' by removing as many pairs $v_i w_i$ and $v_j w_j$ of pendant edges as possible until there is at most one pendant edge left, and replacing these pairs by edges $v_i v_j$. Then G'' may contain some double edges but no triple edges. By Vizing's theorem, this graph can be edge-coloured with $t - n$ colours and then this edge-colouring may be equalized in such a way that, if there is a solitary pendant edge e, then no colour class contains more edges than the colour class in which e lies. This edge-colouring induces a corresponding edge colouring of G' which satisfies the conditions of Theorem 4.4, so R can be embedded in a symmetric latin square of side t.

5. EMBEDDING PARTIAL (p, q, x) - LATIN RECTANGLES.

An $r \times s$ *partial* (p, q, x) - *latin rectangle* on symbols $\sigma_1, \ldots, \sigma_n$ is an $r \times s$ matrix in which each cell contains at most x symbols (counting repetitions), each symbol occurring at most p times in each row and at most q times in each column. An $r \times s$ *incomplete* (p, q, x) - *latin rectangle* on symbols $\sigma_1, \ldots, \sigma_n$ is an $r \times s$ partial (p, q, x) - latin rectangle in which each cell contains precisely x symbols (counting repetitions). If each symbol occurs exactly p times in each row and exactly q times in each column and each cell contains exactly x symbols (counting repetitions) then the (p, q, x) - latin rectangle is called *exact* and in that case it is shown in [2] that, for some positive rational number t, $r = pt$, $s = qt$ and $n = xt$. We shall assume throughout that t has the property that xt, pt and qt are all integers. We may require that no symbol in any cell is repeated, in which case the (p, q, x) - latin rectangle is said to be *without repetition*; otherwise we call it a (p, q, x)- latin rectangle, *repetition permitted*. Please note that the terminology here is a slight variation of that in [2].

An $r \times s$ partial (p, q, x)- latin rectangle, repetition permitted, on symbols $\sigma_1, \ldots, \sigma_n$ is *saturated* if, for each cell c containing fewer than x symbols (counting repetitions), each of the symbols $\sigma_1, \ldots, \sigma_n$ either occurs p times in the row in which c lies or occurs q times in the column in which c lies.

An $r \times s$ partial (p, q, x)- latin rectangle without repetition on symbols $\sigma_1, \ldots, \sigma_n$ is *saturated* if, for each cell c containing fewer than x symbols, each of the symbols $\sigma_1, \ldots, \sigma_n$ either occurs in the cell c, or occurs p times in the row in which c

lies or occurs q times in the column in which c lies.

First we consider the problem of embedding $r \times s$ partial (p, q, x)- latin rectangles, repetition permitted. The following theorem [3] is needed.

<u>Theorem 5.1.</u> An $r \times s$ incomplete (p, q, x)- latin rectangle R, repetition permitted, on symbols $\sigma_1, \ldots, \sigma_{xt}$ can be embedded in an exact (p, q, x)- latin rectangle, repetition permitted, on the same symbols if and only if

$$N_R(\sigma_i) \geq qs + pr - pqt \qquad (1 \leq i \leq t).$$

Let the deficiency of the cell in row ρ and column c be x minus the number of symbols (counting repetitions) in the cell; let it be denoted by $D_{\rho c}$. Let $D_\rho = \sum_{c=1}^{s} D_{\rho c}$, $D_c = \sum_{\rho=1}^{r} D_{\rho c}$ and $D_R = \sum_{\rho=1}^{r} D_\rho$. We now give the following generalization of Theorem 1.2.

<u>Theorem 5.2.</u> A saturated partial (p, q, x)- latin rectangle R (repetition permitted) of size $r \times s$ on n symbols $\sigma_1, \ldots, \sigma_n$ can be embedded in an exact (p, q, x)- latin rectangle (repetition permitted) on xt symbols $\sigma_1, \ldots, \sigma_{xt}$ if and only if

(i) $D_R \geq (qs+pr-pqt)(xt-n)$,

(ii) $N_R(\sigma_i) \geq qs + pr - pqt \qquad (1 \leq i \leq n)$,

(iii) $D_\rho \leq p(xt-n)$ and $D_c \leq q(xt-n) \qquad (1 \leq \rho \leq r, 1 \leq c \leq s)$.

We omit the proof as it is very similar to that of Theorems 1.2. and 5.4.

We next consider the problem of embedding $r \times s$ partial (p, q, x)- latin rectangles without repetition. We use the following theorem [3].

<u>Theorem 5.3.</u> An incomplete $r \times s$ (p, q, x)- latin rectangle R without repetition on symbols $\sigma_1, \ldots, \sigma_{xt}$ can be embedded in an exact (p, q, x)- latin rectangle without repetition on the same symbols if and only if the following four conditions are obeyed:

(i) $N_R(\sigma_i) \geq qs + pr - pqt$ $(1 \leq i \leq xt)$,

(ii) $N_R(\sigma_i) \leq qs + pr - pqt + (pt-s)(qt-r)$ $(1 \leq i \leq xt)$,

(iii) $N_\rho(\sigma_i) \geq p + s - pt$ for all rows ρ of R $(1 \leq i \leq xt)$,

(iv) $N_c(\sigma_i) \geq q + r - qt$ for all columns c of R $(1 \leq i \leq xt)$.

Here $N_\rho(\sigma_i)$ and $N_c(\sigma_i)$ are the number of times σ_i occurs in row ρ and column c respectively. The result we deduce is the following.

<u>Theorem 5.4</u>. A saturated partial (p, q, x)-latin rectangle R without repetition of size $r \times s$ on n symbols $\sigma_1, \ldots, \sigma_n$ can be embedded in an exact (p, q, x)-latin rectangle T without repetition on xt symbols $\sigma_1, \ldots, \sigma_{xt}$ if and only if the following eight conditions are obeyed:

(i) $D_R \geq (qs+pr-pqt)(xt-n)$,

(ii) $D_R \leq (qs+pr-pqt+(pt-s)(qt-r))(xt-n)$,

(iii) $N_R(\sigma_i) \geq qs + pr - pqt$ $(1 \leq i \leq n)$,

(iv) $N_R(\sigma_i) \leq qs + pr - pqt + (pt-s)(qt-r)$ $(1 \leq i \leq n)$,

(v) $D_\rho \leq p(xt-n)$ ($\forall$ rows ρ of R),

and $D_c \leq q(xt-n)$ ($\forall$ columns c of R),

(vi) $D_\rho \geq (p+s-pt)(xt-n)$ ($\forall$ rows ρ of R),

and $D_c \geq (q+r-qt)(xt-n)$ ($\forall$ columns c of R),

(vii) $N_\rho(\sigma_i) \geq p + s - pt$ ($\forall$ rows ρ of R, $1 \leq i \leq n$)

and $N_c(\sigma_i) \geq q + r - qt$ ($\forall$ columns c of R, $1 \leq i \leq n$),

(viii) $D_{\rho c} \leq xt - n$ ($\forall$ rows ρ and columns c of R).

1. <u>Necessity</u>.

(i), (iii) and (v) are just slight variations of the conditions in Theorem 1.2.

(iii), (iv) and (vii). Since R is saturated, symbols $\sigma_1, \ldots, \sigma_n$ must satisfy the Ryser-type conditions of Theorem 5.3.

(ii) and (iv). Each of $\sigma_{n+1}, \ldots, \sigma_{xt}$ must satisfy the Ryser-type conditions of Theorem 5.3, so the deficiences D_R, D_ρ and D_c must be sufficiently large for this to be possible.

(viii). The deficiency of each cell has to be filled with the symbols $\sigma_{n+1}, \ldots, \sigma_{xt}$, none being used more than once in any cell.

2. **Sufficiency.** Suppose conditions (i)-(viii) all hold. Consider the bipartite graph G on vertices $\rho_1, \ldots, \rho_r, c_1, \ldots, c_s$ formed by joining ρ_i to c_j by k edges if the deficiency of the cell (i, j) is k. Then, by (i) and (ii),

$$(qs+pr-pqt + (pt-s)(qt-r))(xt-n) \geq |E(G)|$$
$$\geq (qs+pr-pqt)(xt-n)$$

and, by (v) and (vi),

$$(p+s-pt)(xt-n) \leq D_\rho \leq p(xt-n) \qquad (\forall \text{ rows } \rho \text{ of } R)$$

and

$$(q+r-qt)(xt-n) \leq D_c \leq q(xt-n) \qquad (\forall \text{ columns } c \text{ of } R).$$

By de Werra's theorem, G has a balanced edge-colouring with $xt-n$ colours $\sigma'_{n+1}, \ldots, \sigma'_{xt}$ and this balanced colouring can be equalized. Then the number of edges of each colour will at least $qs + pr - pqt$ and at most $qs + pr - pqt + (pt-s)(qt-r)$; furthermore, at each vertex ρ_i the number of edges of each colour will be at least $p + s - pt$ and at most p, and at each vertex c_j the number of edges of each colour will be at least $q + r - qt$ and at most q.

We also note that the number k of edges joining ρ_i to c_j is the deficiency of the cell (i, j) and so, by (viii), is at most $xt - n$, so each parallel edge of a multiedge receives a different colour.

We then put σ_ℓ into cell (i, j) whenever there is an edge joining ρ_i and c_j coloured σ'_ℓ. Then the Ryser-type conditions of Theorem 5.3 will be obeyed by all the symbols and there will be x symbols in each cell, at most p occurrences of each symbol in each row and at most q occurrences of each symbol in each column. Then, by Theorem 5.3 R can be embedded in an exact (p, q, x)- latin rectangle without repetition on symbols $\sigma_1, \ldots, \sigma_{xt}$.

6. SYMMETRIC (p, p, x)- LATIN SQUARES.

If p is odd then the theorems one could obtain concerning embedding partial symmetric (p, p, x)-squares would be generalizations of Theorems 4.2 and 4.4 and their corollaries, Corollaries 4.3 and 4.5. Thus the theorems would be in terms of graphs; numerical conditions by themselves would be sufficient but not necessary for the embedding. As the results to be expected are therefore rather weak, we do not give them here. However, if p is even, then we have quite a different story, and we obtain numerical conditions which are necessary and sufficient for the embedding.

First we consider symmetric (p, p, x)- latin squares, repetition permitted. We shall need the following theorem [3].

Theorem 6.1. Let p be even. An incomplete symmetric (p, p, x)- latin square R of size $r \times r$ on symbols $\sigma_1, \ldots, \sigma_{xt}$ can be embedded in an exact symmetric (p, p, x)- latin square, repetition permitted, if and only if

(i) $\quad N_R(\sigma_i) \geq 2pr - p^2 t \qquad (1 \leq i \leq xt)$,

(ii) there are at most $x(pt-r)$ values of i, $1 \leq i \leq xt$, for which $N_R(\sigma_i)$ is odd.

From this we deduce the following result about partial symmetric (p, p, x)- latin squares.

Theorem 6.2. Let p be even. A saturated symmetric partial (p, p, x)- latin square R of size $r \times r$ on n symbols $\sigma_1, \ldots, \sigma_n$ can be embedded in an exact symmetric (p, p, x)- latin square T on the symbols $\sigma_1, \ldots, \sigma_{xt}$ of size $t \times t$ with repetition permitted if and only if the following four conditions hold:

(i) $\quad D_R \geq (2pr - p^2 t)(xt-n)$,

(ii) $\quad N_R(\sigma_i) \geq 2pr - p^2 t \qquad (1 \leq i \leq n)$,

(iii) $\quad D_\rho \leq p(xt-n) \qquad (\forall \text{ rows } \rho \text{ of } R)$,

(iv) $\quad y \leq \begin{cases} x(pt-r) & \text{if } D_{diag} \text{ is even} \\ x(pt-r)-1 & \text{if } D_{diag} \text{ is odd.} \end{cases}$

Here y_R is the number of values of i, $1 \leq i \leq n$, such that σ_i occurs an odd number of times on the diagonal of R and
$$D_{diag} = \sum_{\rho=1}^{r} D_{\rho\rho}.$$

Proof.

1. Necessity.

(i), (ii) and (iii). The proofs of these are as in the proofs of Theorem 1.2.

(iv). If z is the number of symbols σ_i with $n+1 \le i \le xt$ which occur on the diagonal of R an odd number of times then, by Theorem 6.1 and since R is saturated, $y+z \le x(pt+r)$. If D_{diag} is even then z could be 0, but if D_{diag} is odd then z must be at least one.

Sufficiency. Suppose (i), (ii), (iii) and (iv) hold. Let G be the multigraph on vertices $v_1,\ldots,v_n$, as well as some further vertices, formed as follows. Vertices v_i and v_j are joined by k edges if cells (i, j) and (j, i) each have deficiency k; to vertex v_i adjoin D_{ii} pendant edges. Now form a multigraph G' by pairing the pendant edges off as much as possible, so that there is at most one pendant edge left over, and replacing each pair $v_i w_\alpha$, $v_j w_\beta$ of pendant edges by an edge $v_i v_j$. There will be one pendant edge left over if and only if D_{diag} is odd. Then the maximum degree of G is $\le p(xt-n)$ (by (iii)) (each loop counting two towards the degree) and the number of edges of G is at least $\frac{1}{2}(2pr-p^2t)(xt-n)$ (by (i)). Since p is even, the edges of G' can be coloured with $xt-n$ colours $\sigma'_{n+1},\ldots,\sigma'_{xt}$ so that not more than p edges of each colour appear at any vertex ([10], [18]). Then this colouring can be equalized, whilst retaining this latter property. Then we place σ_k in cells (i, j) and (j, i) $(j \ne i)$ ℓ times each if ℓ edges joining v_i and v_j are coloured σ'_k, and σ_k is placed in cell (i, i) ℓ times if there are ℓ pendant edges of G (in the edge colouring of G corresponding to that of G') coloured σ'_k. Then no symbol appears more than p times in any row and each symbol appears at least $2pr - p^2t$ times altogether in R. Furthermore the number of symbols from $\sigma_{n+1},\ldots\sigma_{xt}$ which appear on the diagonal of R an odd number of times is

$$\begin{cases} 0 & \text{if } D_{diag} \text{ is even,} \\ 1 & \text{if } D_{diag} \text{ is odd;} \end{cases}$$

so the number of symbols appearing an odd number of times on the diagonal of R is at most $x(pt-r)$. By Theorem 6.1, R can now be embedded in an exact symmetric (p, p, x)-latin square on the symbols $\sigma_1,\ldots,\sigma_{xt}$ of size $t \times t$, repetition permitted.

If the symmetric (p, p, x)-latin square T is to be without repetition, then there does not seem to be a completely satisfactory analogue of Theorem 6.2. The trouble seems to lie in the fact that there are multigraphs of maximum degree p (p even) which have equitable edge-colourings with p/2 colours, but do not have balanced edge-colourings with p/2 colours. An example is given in Figure 1.

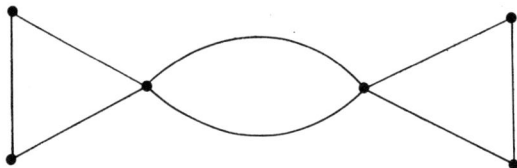

Figure 1. A multigraph with an equitable, but without a balanced, edge-colouring with two colours.

There is good reason to suppose that there is a good analogue of Theorem 6.2 in the case when the symmetric (p, p, x)- latin squares are permitted to have symbols occurring up to two times in a cell, but we have not tried to work out the details of this. The same remark applies to the symmetric blocked (p, p, x)- latin squares we are about to discuss.

A *partial symmetric blocked* (p, p, x)- *latin square* on the symbols $\sigma_1, \ldots, \sigma_n$ is a partial symmetric (p, p, x)- latin square on those symbols in which the diagonal cells are all empty. It is *saturated* if to each nondiagonal cell with positive deficiency each symbol either occurs p times in the row in which c lies or the column. An *incomplete symmetric blocked* (p, p, x)-*latin square* on symbols $\sigma_1, \ldots, \sigma_n$ is a partial symmetric blocked (p, p, x)- latin square in which each non-diagonal cell contains x symbols exactly (counting repetitions). An *exact symmetric blocked* (p, p, x)- *latin square* on symbols $\sigma_1, \ldots, \sigma_n$ is an incomplete symmetric blocked (p, p, x)- latin square on those symbols in which each symbol occurs exactly p times in each row. The existence of an exact symmetric blocked (p, p, x)- latin square implies the existence of a positive rational number t such that n = xt and the number of rows is pt + 1; conversely if t is a positive rational such that xt and pt are positive integers then there is a (pt+1) × (pt+1) symmetric blocked (p, p, x)- latin square on xt symbols (see [2]). As before we shall always assume that xt and pt are positive integers. Exact symmetric blocked (p, p, x)- latin squares have a natural interpretation as edge-colourings of multigraphs in which each pair of vertices are joined by precisely x edges; in particular exact symmetric blocked (p, p, 1)- latin squares correspond to types of edge-colourings of complete graphs.

The following theorem from [3] is required.

Theorem 6.3. An incomplete symmetric blocked (p, p, x)- latin square R of size $r \times r$ on symbols $\sigma_1, \ldots, \sigma_{xt}$ can be embedded in an exact symmetric blocked (p, p, x)- latin square, repetition permitted, on the same symbols if and only if

(i) $N_R(\sigma_i) \geq 2pr - p(pt+1)$ $(1 \leq i \leq xt)$,

(ii) $p(pt+1)$ is even.

From this the following theorem may be deduced; we omit the proof as it is so similar to that of Theorem 6.2.

Theorem 6.4. Let p be even. A saturated blocked symmetric partial (p, p, x)- latin square on n symbols $\sigma_1, \ldots, \sigma_n$ can be embedded in a symmetric blocked $(t+1) \times (t+1)$ (p, p, 1)- latin square T on xt symbols $\sigma_1, \ldots, \sigma_{xt}$ if and only if

(i) $D_R \geq (2pr-p(pt+1))(xt-n)$,

(ii) $N_R(\sigma_i) \geq 2pr - p(pt+1)$ $(1 \leq i \leq n)$,

(iii) $D_\rho \leq p(xt-n)$ ($\forall$ rows ρ of R).

For a partial symmetric blocked (p, p, 1) latin square R of side $r \times r$, let G_R be the graph with vertices $v_1, \ldots, v_r$ formed by joining v_i and v_j if cells (i, j) and (j, i) are empty. We have the following theorem about partial symmetric blocked (p, p, 1)- latin squares.

Theorem 6.5. Let p be odd. A saturated blocked symmetric (p, p, 1)- partial latin square on n symbols $\sigma_1, \ldots, \sigma_n$ can be embedded in an exact symmetric blocked $(t+1) \times (t+1)$ (p, p, 1)- latin square T, repetition permitted, on t symbols $\sigma_1, \ldots, \sigma_t$ if and only if

(i)   t is odd,

(ii) $D_R \geq (2pr-p(pt+1))(t-n)$,

(iii) $N_R(\sigma_i) \geq 2pr - p(pt+1)$ $(1 \leq i \leq n)$,

(iv)   G can be edge-coloured with t - n colours so that

(a) no colour appears on more than p edges at any point
(b) each colour appears on at least $\frac{1}{2}(2pr-p(pt+1))$ edges.

We omit the proof as it is similar to others in this section. The fact that t is odd comes from the fact that $pt + 1$ must be even in Theorem 6.3.

Corollary 6.6. Let p be odd. A blocked symmetric $(p, p, 1)$-partial latin square on n symbols $\sigma_1,\ldots,\sigma_n$ can be embedded in a symmetric blocked $(t+1) \times (t+1)$ $(p, p, 1)$-latin square T on t symbols $\sigma_1,\ldots,\sigma_t$ if (i), (ii) and (iii) of Theorem 6.5 hold and the number of vacant cells (excluding the diagonal cell) in each row is $< p(t-n)$.

Proof. By Vizing's theorem for a simple graph G, if $\Delta(G) < p(t-n)$ then G can be edge-coloured with $t - n$ colours so that not more than p edges at any vertex receive the same colour. Then this edge-colouring can be equalized. Then condition (iv) of Theorem 6.5 will be satisfied so R can be embedded in an exact symmetric blocked $(t+1) \times (t+1)$ $(p, p, 1)$-latin square T, repetition permitted.

7. EMBEDDING WITH SPECIFIED DIAGONALS.

In this section we give a theorem about embedding a partial $r \times s$ latin rectangle R in a latin square T in which some of the cells of T outside R are also specified.

We use two functions

$$f : \{\sigma_1,\ldots,\sigma_t\} \to \{0, 1\} \quad \text{and} \quad g : \{\sigma_1,\ldots,\sigma_t\} \to \{0, 1,\ldots,t\}.$$

If $r > s$ we call the set $\{(r+1, s+1), (r+2, s+2),\ldots,(n, n-r+s)\}$ of cells, the $(r-s)$-th lower diagonal outside R.

Theorem 7.1. Let $r \geq s \geq 1$, $\sum_{i=1}^{t} g(\sigma_i) \leq \begin{cases} t - r - 1 & \text{if } r = s, \\ t - r & \text{if } r \neq s \end{cases}$
and let R be an incomplete $r \times s$ latin square on the symbols $\sigma_1,\ldots,\sigma_t$. Then R can be embedded in a latin square T on $\sigma_1,\ldots,\sigma_t$ with the symbol σ_i occurring at least $g(\sigma_i)$ times on the $(r-s)$-th lower diagonal outside R $(1 \leq i \leq t)$ if and only if

$$N_R(\sigma_i) \geq r + s - t + g(\sigma_i) \qquad (1 \leq i \leq t).$$

If $r = s$, in the inequality $\sum_{i=1}^{t} g(\sigma_i) \leq t - r - 1$, one cannot replace the bound by $t - r$ — for a discussion of this see [1].

From Theorem 7.1 follows

<u>Theorem 7.2.</u> Let $r \geq s \geq 1$, let $\sum_{i=1}^{t} f(\sigma_i) \leq \begin{cases} t - r - 1 & \text{if } r = s, \\ t - r & \text{if } r \neq s \end{cases}$ and let R be a saturated partial $r \times r$ latin square on the symbols $\sigma_1, \ldots, \sigma_n$. Then R can be embedded in a latin square T on the symbols $\sigma_1, \ldots, \sigma_t$ in which symbol σ_i occurs at least $f(\sigma_i)$ times on the $(r-s)$-th lower diagonal outside R if and only if the following three conditions hold:

(i) $D_R \geq (r+s-t)(t-n) + f(\sigma_{n+1}) + \ldots + f(\sigma_t)$,

(ii) $N_R(\sigma_i) \geq r + s - t + f(\sigma_i)$ $\quad (1 \leq i \leq n)$,

(iii) $D_\rho \leq t - n$ and $D_c \leq t - n$ $\quad (\forall$ rows ρ of R and columns c of $R)$.

The proof of this is completely straightforward following the proof of Theorem 1.2. An interesting question is: to which functions g can Theorem 7.2 be generalized? The generalization to any function g such that

$$\sum_{i=1}^{t} g(\sigma_i) \leq \begin{cases} t - r - 1 & \text{if } r \neq s \\ t - r & \text{if } r = s \end{cases}$$

does not seem to be valid; one would need to properly edge-colour a bipartite graph G with $t - r$ colours so that the i-th colour class has at least $r + s - t + g(\sigma_i)$ colours. For a discussion of this colouring problem see [7], [15] and [19].

REFERENCES

[1] L.D. Andersen, Latin squares and their generalizations, Ph.D. Thesis, Reading, 1979.

[2] L.D. Andersen and A.J.W. Hilton, Generalized latin rectangles I: construction and decomposition, Discrete Math. 31 (1980) 125-152.

[3] L.D. Andersen and A.J.W. Hilton, Generalized Latin rectangles II: embedding, Discrete Math. 31 (1980) 235-260.

[4] L.D. Andersen and A.J.W. Hilton, Thank Evans!, to appear.

[5] L.D. Andersen, R. Häggkvist, A.J.W. Hilton and W. Poucher, Embedding incomplete latin squares in latin squares whose diagonal is almost completely prescribed, Europ. J. Combinatorics 1 (1980) 5-7.

[6] A. Cruse, On embedding incomplete symmetric latin squares, J. Combinatorial Theory (A), 16 (1974) 18-22.

[7] A. Cruse, On extending incomplete latin rectangles, Proc. 5th Southeastern Conf. on Combinatorics, Graph Theory and Computing, Florida Atlantic University, Boca Raton, Florida (1974) 333-348.

[8] J. Folkman and D.R. Fulkerson, Edge-colourings in bipartite graphs, Combinatorial Mathematics and its Applications (Univ. North Carolina Press, Chapel Hill, 1969).

[9] D. König, Theorie der Endlichen und Unendlichen Graphen (Chelsea, New York, 1950).

[10] C.J.H. McDiarmid, The solution of a time-tabling problem, J. Inst. Maths. Applics. 9 (1972) 23-34.

[11] J. Petersen, Die Theorie der regulären Graphen, Acta Math. 15 (1981) 193-220.

[12] H.J. Ryser, A combinatorial theorem with an application to latin squares, Proc. Amer. Math. Soc. 2 (1951) 550-552.

[13] V.G. Vizing, On an estimate of the chromatic class of a p-graph, Diskret. Analiz. 3 (1960) 25-30.

[14] D. de Werra, Balanced schedules, Information J. 9 (1971) 230-237.

[15] D. de Werra, Investigations on an edge-colouring problem, Discrete Math. 1 (1971) 167-179.

[16] D. de Werra, Equitable colorations of graphs, Rev. Franc. Rech. Oper. 5 (1971) 3-8.

[17] D. de Werra, A few remarks on chromatic scheduling, in: B. Roy, ed., Combinatorial Programming: Methods and Applications (D. Reidel Publ. Co., Dordrecht-Holland, 1975) 337-342.

[18] D. de Werra, On a particular conference scheduling problem, Information J. 13 (1975) 308-315.

[19] D. de Werra, Multigraphs with quasiweak odd cycles, J. Combinatorial Theory (B), 23 (1977) 75-82.

[20] D. de Werra, Obstructions for regular colourings, to appear.

EDGE-COLOURING REGULAR BIPARTITE GRAPHS

A.J.W. HILTON and C.A. RODGER

University of Reading
England

We give necessary and sufficient conditions for
a regular bipartite multigraph of degree m to
have an edge-colouring with m colours in which
two specified edges receive the same colour.
Call a cubic bipartite multigraph G with bipartition (A, B) *skew* if there is a cutset K
of six edges whose removal separates G into two
subgraphs G_1 and G_2 such that each edge of K
joins a vertex of A ∩ $V(G_1)$ to a vertex of
B ∩ $V(G_2)$. We consider the problem of finding
necessary and sufficient conditions for a skew
cubic bipartite multigraph to have an edge-
colouring with three colours in which three
specified edges of K receive different colours.
This problem arose in an investigation concerning
latin squares. We conclude with a conjecture on
this problem.

If G is a multigraph, if A and B are disjoint sets of vertices of G and if K is a set of edges of G, then we let V(G) and E(G) denote the vertex set and edge set respectively of G, we let E(A, B) denote the set of edges joining vertices in A to vertices in B and we let N(K) denote the set of vertices of G which are incident with at least one edge of K.

An *edge-colouring* of G is an assignment of colours to the edges of G so that no two edges of the same colour have a common vertex. König [5] proved that a bipartite multigraph with maximum degree m can be edge-coloured with m colours.

In the first part of this paper we consider some edge-colouring results for regular bipartite multigraphs of general degree $m \geq 2$.

In the second part we consider some more specific problems concerning skew cubic bipartite multigraphs. In the third part we state a conjecture for which our results provide strong evidence.

All the results have the common theme that the graph has to be edge-coloured with a given number of colours with certain specified edges receiving specified colours.

1. REGULAR BIPARTITE MULTIGRAPHS OF DEGREE m.

Our first object is to prove the following theorem.

THEOREM 1. Let G be a regular bipartite multigraph of degree m. Let e_1 and e_2 be two edges of G. Then G can be edge-coloured with m colours with e_1 and e_2 receiving the same colour if and only if G does not contain a cutset K containing e_1 and e_2 with $|K| = m$ which separates G into two disjoint submultigraphs G_1 and G_2 such that each edge of K joins a vertex of $A \cap V(G_2)$ to a vertex of $B \cap V(G_1)$ for some bipartition (A, B) of G. The exceptional multigraph of Theorem 1 is illustrated in Figure 1.

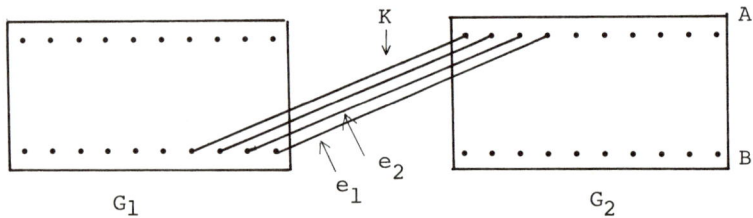

Figure 1. The exceptional multigraph of Theorem 1.

In the proof of Theorem 1, we make use of the following lemma due to Häggkvist [3].

Lemma (Häggkvist). Let G be a regular bipartite multigraph with bipartition (A, B), let $e \in E(G)$ and let $E^* \subseteq E(G) \setminus \{e\}$. Then G has a 1-factor which contains e and avoids E^* if and only if there do not exist disjoint sets E_1 and E_2 of edges such that $e \in E_1$, $E_2 \subseteq E^*$, $|E_1| = |E_2|$ and $E_1 \cup E_2$ is a cutset whose removal separates G into two disjoint subgraphs G_1 and G_2 such that $E_1 \subseteq E(A \cap V(G_1), B \cap V(G_2))$ and $E_2 \subseteq E(A \cap V(G_2), B \cap V(G_1))$.

The exceptional multigraph of Häggkvist's lemma is illustrated in Figure 2. In fact Häggkvist did not make it clear that his lemma applies to multigraphs, but an examination of his proof shows that it does.

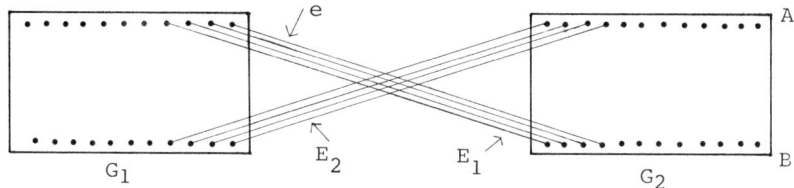

Figure 2. The exceptional multigraph of Häggkvist's lemma.

Proof of Theorem 1.

1. Necessity. Suppose that G can be edge-coloured with m colours so that two specified edges, e_1 and e_2, receive the same colour. Suppose that G contains a cutset K with $|K| = m$ and $e_1 \in K$ which separates G into two disjoint subgraphs G_1 and G_2 with each edge of K joining a vertex of $A \cap V(G_2)$ to a vertex of $B \cap V(G_1)$. Each vertex of $B \cap V(G_2)$ is joined by m edges to vertices of $A \cap V(G_2)$, but the m edges of K join vertices of $A \cap V(G_2)$ to vertices which are not in $B \cap V(G_2)$. Therefore $|B \cap V(G_2)| = |A \cap V(G_2)| - 1$. Therefore each colour has to be used on exactly one edge of K, and so $e_2 \notin K$.

2. Sufficiency. Suppose that G does not contain a cutset K with $|K| = m$ and $e_1, e_2 \in K$ which separates G into two disjoint subgraphs G_1 and G_2 such that each edge of K joins a vertex of $A \cap V(G_2)$ to a vertex of $B \cap V(G_1)$ for some bipartition (A, B). Let e_2 be incident with a vertex v in A and let E^* be the set of all the other edges incident with v. Then $e_1 \notin E^*$ for otherwise $E^* \cup \{e_2\}$ would be a set K of the forbidden type with $V(G_2) = \{v\}$.

If there is a 1-factor F containing e_1 and avoiding E^* then F contains e_2 also. Then $G \setminus F$ is a regular bipartite graph of degree $m - 1$ and so, by König's theorem, can be edge-coloured with $m - 1$ colours. Thus G can be edge-coloured with m colours with e_1 and e_2 receiving the same colour.

If there is no 1-factor containing e_1 and avoiding E^* then, by Häggkvist's lemma, there are sets E_1 and E_2 of edges such that $e_1 \in E_1$, $E_2 \subseteq E^*$, $|E_1| = |E_2|$ and $E_1 \cup E_2$ is a cutset which separates G into two subgraphs G_1^* and G_2^* such that the edges of E_1 join vertices of $V(G_1^*) \cap A$ and $V(G_2^*) \cap B$ and the edges of E_2 join vertices of $V(G_1^*) \cap B$ and $V(G_2^*) \cap A$. Let $E_3 = (E^* \cup \{e_2\}) \setminus E_2$. Then $E_2 \cup E_3$ is the set of edges incident with v. The situation is illustrated in Figure 3.

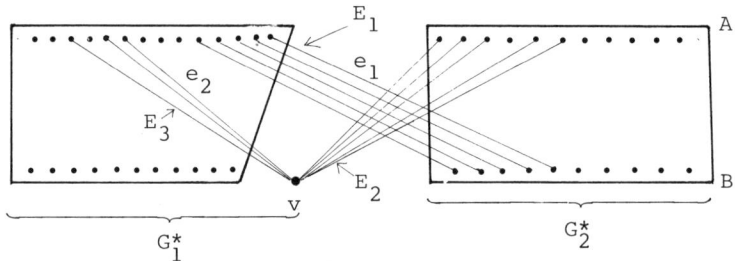

Figure 3.

Let $G_1 = G_1^* \setminus \{v\}$ and let G_2 be G_2^* with the vertex v and the edges of E_2 adjoined. Then the edges of $E_1 \cup E_3$, which number $|E_1| + (m - |E_2|) = m$ and include e_1 and e_2, join $A \cap V(G_1)$ to $B \cap V(G_2)$ and no edges join $A \cap V(G_2)$ to $B \cap V(G_1)$. But then these m edges form a cutset K of the forbidden type, a contradiction. This proves Theorem 1.

We also deduce two more specialized consequences of Häggkvist's lemma, of which use will be made later. Call a regular bipartite multigraph of degree $m \geq 2$ a *circle multigraph* if it has the following form. For some bipartition (A, B), G has a cutset F with $|F| = m$ whose removal separates G into two disjoint submultigraphs G_1 and G_2 such that each edge of F joins $B \cap V(G_1)$ to $A \cap V(G_2)$; G also has a second cutset $\{e_1, e_2\}$, where $\{e_1, e_2\} \cap F = \phi$, which separates G into two disjoint submultigraphs H_1 and H_2; $|E(H_1) \cap F| = 1$, $|E(H_2) \cap F| = m - 1$, e_1 joins a vertex of $A \cap V(H_1) \cap V(G_1)$ to a vertex of $B \cap V(H_2) \cap V(G_1)$ and e_2 joins a vertex of $B \cap V(H_1) \cap V(G_2)$ to a vertex of $A \cap V(H_2) \cap V(G_2)$. A circle multigraph is illustrated in Figure 4.

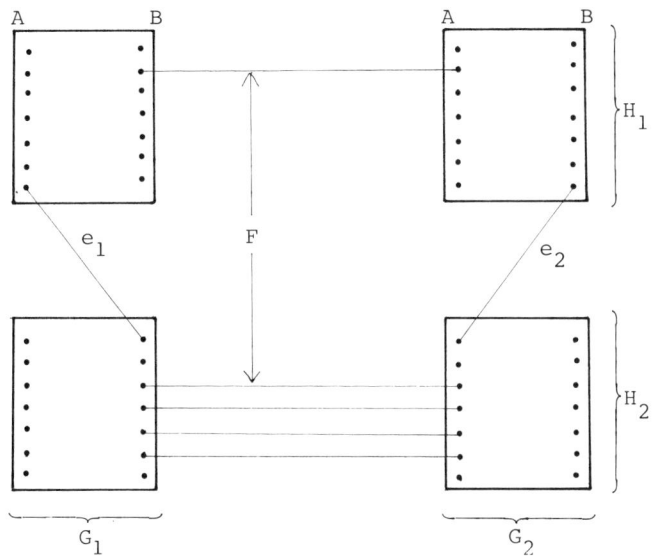

Figure 4. A circle multigraph.

THEOREM 2. Let G be a regular bipartite multigraph of degree m with a cutset F with the properties that $|F| = m$ and the removal of F separates G into two disjoint submultigraphs G_1 and G_2 such that, for some bipartition (A, B), each edge of F joins a vertex of $A \cap V(G_1)$ to a vertex of $B \cap V(G_2)$. Let $e_1 \in E(G_1)$ and $e_2 \in E(G_2)$. Then G can be edge-coloured with m colours with e_1, e_2 receiving different colours if and only if G is not a circle multigraph.

Proof. By Häggkvist's lemma, G has an edge-colouring with m colours so that e_1 and e_2 receive different colours if and only if $\{e_1, e_2\}$ is not a cutset. If $\{e_1, e_2\}$ is a cutset then it is easy to see by counting arguments that G is a circle multigraph.

Now call a regular bipartite multigraph of degree $m \geq 3$ a *double-circle multigraph* if it has the following form. For some bipartition (A, B), G has a cutset F with $|F| = m$ whose removal separates G into two disjoint submultigraphs G_1 and G_2 such that each edge of F joins $B \cap V(G_1)$ to $A \cap V(G_2)$; G also has a second cutset $E_1 \cup E_2$, where $E_1 \subset E(G_1)$, $E_2 \subset E(G_2)$ and $|E_1| = |E_2| = 2$, which separates G into two disjoint subgraphs

H_1 and H_2; $|E(H_1) \cap F| = 2$, $|E(H_2) \cap F| = m - 2$, the edges of E_1 join vertices of $A \cap V(H_1) \cap V(G_1)$ to vertices of $B \cap V(H_2) \cap V(G_1)$ and the edges of E_2 join vertices of $B \cap V(H_1) \cap V(G_2)$ to vertices of $A \cap V(H_2) \cap V(G_2)$. A double-circle multigraph would be illustrated by Figure 4 if each of the single edges were replaced by two edges.

THEOREM 3. Let G be a regular bipartite multigraph of degree m with a cutset F with the properties that $|F| = m$ and the removal of F separates G into two disjoint submultigraphs G_1 and G_2 such that, for some bipartition (A, B), each edge of F joins a vertex of $A \cap V(G_1)$ to a vertex of $B \cap V(G_2)$. Let G not contain a cutset of two edges, one of which is a member of F. Let $e_1 \in E(G_1)$, $e_2 \in E(G_2)$ and $f \in F$. Then G can be edge-coloured with m colours with the edges e_1, e_2, f receiving different colours if and only if G is not a double-circle multigraph with $e_1 \in E_1$, $e_2 \in E_2$ and $f \in F \cap E(H_1)$.

Proof. The necessity is obvious. We shall prove the sufficiency.

Suppose that G can be edge-coloured with m colours with e_1, e_2 and f receiving different colours. Consider two regular bipartite multigraphs G_1^{**} and G_2^{**} of degree m constructed as follows. Let G_1^* and G_2^* be the multigraphs induced by $V(G_1) \cup (A \cap V(G_2))$ and $V(G_2) \cup (B \cap V(G_1))$ respectively and let G_1^{**} and G_2^{**} be formed from G_1^* and G_2^* by identifying the vertices of $A \cap V(G_2)$ and $B \cap V(G_1)$ respectively. Let the edges of F be $f = f_1, f_2, \ldots, f_m$.

By Häggkvist's lemma and König's theorem, for any i, $1 \le i \le m$, G_1^{**} and G_2^{**} can be edge-coloured with m colours so that e_1 and f_i or e_2 and f_i respectively receive different colours. Suppose that there is some i, $2 \le i \le m$, such that either G_1^{**} or G_2^{**} (say G_2^{**} to be specific) has no 1-factor including e_1 or e_2 respectively (say e_2) and avoiding f_1 and f_i. Then by Häggkvist's lemma, G_2^{**} has a cutset of four edges including e_2, f_1 and f_i and say e which separates G_2^{**} into two submultigraphs M_1 and M_2, where f_1 and f_i join vertices of $B \cap V(M_1)$ to vertices of $A \cap V(M_2)$ and e and e_2 join vertices of $A \cap V(M_1)$ to vertices of $B \cap V(M_2)$. Also in every edge-colouring of G in which e_2 and f_1 receive different colours, e_2 and f_i receive the same colour.

Consequently, for G_1^{**} there is an edge-colouring with e_1, f_1 and f_i receiving different colours. Therefore, by Häggkvist's lemma, G_1^{**} cannot have a cutset consisting of four edges including e_1, f_1 and f_i, so G cannot be a double-circuit multigraph with $e_1 \in E_1$, $e_2 \in E_2$ and $f \in F \cap E(H_1)$.

2. SKEW CUBIC BIPARTITE MULTIGRAPHS.

Call a cubic bipartite multigraph *skew* if there is a cutset K of six edges whose removal separates G into two disjoint submultigraphs G_1 and G_2 such that each edge of K joins a vertex of $A \cap V(G_1)$ to a vertex of $B \cap V(G_2)$ for some bipartition (A, B), and call K the *skew cutset*. Our second objective, as yet unachieved, is to characterize those skew cubic bipartite multigraphs for which there exists an edge-colouring with three colours such that three specified edges of K each receive different colours. We feel that the results here provide strong evidence for the conjecture concerning this characterization problem we make in the last Section. The problem arose in connection with a possible method of solving a conjecture [4] concerning the embedding of partial idempotent latin squares. For details of considerable progress using this method, see [1]. The embedding problem has since been resolved a different way [2], but the solution of our characterization problem would nevertheless materially aid our understanding of the embedding problem and would tell us which idempotent latin squares of order n can be embedded in idempotent latin squares of order $t \leq 2n$.

We first look at the simpler problem of deciding which skew cubic bipartite multigraphs G possess an edge-colouring with three colours such that two specified edges e_1 and e_2 of K receive different colours. By König's theorem there is such an edge-colouring if and only if G contains a 1-factor containing e_1 and avoiding e_2. From Häggkvist's lemma we see that G contains such a 1-factor if and only if $\{e_1, e_2\}$ does not form a cutset.

At this juncture, we define the exceptional multigraphs of Theorem 4. Call a skew cubic bipartite multigraph with skew cutset K, whose removal separates G into two disjoint submultigraphs G_1 and G_2, a *figure-of-eight* multigraph if there is a cutset $H \subset K$ such that $|H| = 2$ which separates G into two disjoint submultigraphs whose vertex sets are C and D and which has the following properties: Each of G_1 and G_2 contain two disjoint submultigraphs, those of G_1 being induced by $V(G_1) \cap C$ and $V(G_1) \cap D$ and those of G_2 being induced by $V(G_2) \cap C$ and $V(G_2) \cap D$; for some bipartition (A, B) of G, two edges of K join $B \cap C \cap V(G_1)$ to $A \cap C \cap V(G_2)$, two edges of K join $B \cap D \cap V(G_1)$ to $A \cap D \cap V(G_2)$, one edge of K joins $B \cap D \cap V(G_1)$ to $A \cap C \cap V(G_2)$ and one edge of K joins $B \cap C \cap V(G_1)$ to $A \cap D \cap V(G_2)$. A figure-of-eight multigraph is illustrated in Figure 5.

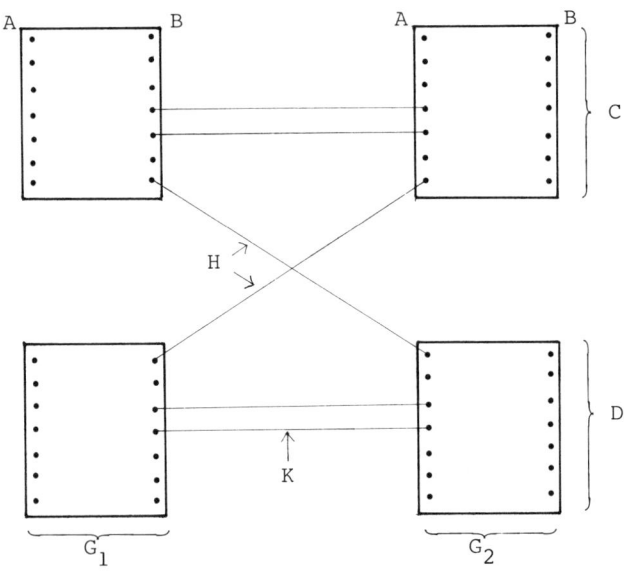

Figure 5. A figure-of-eight multigraph.

THEOREM 4. Let G be a skew cubic bipartite multigraph with skew cutset K. Let $H \subset K$ be a cutset and let $|H| = 2$. Then G is a figure-of-eight multigraph.

Proof. Suppose that H separates G into two disjoint submultigraphs whose vertex sets are C and D. Consider the four submultigraphs of G induced by $C \cap V(G_1)$, $C \cap V(G_2)$, $D \cap V(G_1)$ and $D \cap V(G_2)$. These vertex sets form a partition of $V(G)$. Since H is a cutset, there are no edges between $C \cap V(G_1)$ and $D \cap V(G_1)$

or between $C \cap V(G_2)$ and $D \cap V(G_2)$. Suppose there are $\alpha, \beta, \gamma, \delta$ edges between $C \cap V(G_1)$ and $C \cap V(G_2)$, between $C \cap V(G_1)$ and $D \cap V(G_2)$, between $D \cap V(G_1)$ and $C \cap V(G_2)$ and between $D \cap V(G_1)$ and $D \cap V(G_2)$ respectively. Since the only edges between $V(G_1)$ and $V(G_2)$ are the edges of K we have $\alpha + \beta + \gamma + \delta = 6$. Since $|H| = 2$ we have $\beta + \gamma = 2$. Since G is cubic and the only edges of G joining any of the submultigraphs to the rest of the multigraph are in K, by a simple counting argument we have $\alpha + \beta, \gamma + \delta, \alpha + \gamma, \beta + \delta \equiv 0 \pmod{3}$. It is now an exercise to check that the only solution to this set of equations and congruences is $\beta = \gamma = 1$, $\alpha = \delta = 2$, from which Theorem 4 follows.

The following corollaries now follow easily.

Corollary 5. Let G be a skew cubic bipartite multigraph with skew cutset K. Let e_1 and e_2 be two edges of K. Then G can be edge-coloured with three colours with e_1 and e_2 receiving different colours if and only if G is not a figure-of-eight multigraph with a cutset $\{e_1, e_2\} \subset K$.

Corollary 5. Let G be a skew bipartite multigraph with skew cutset K and with a cutset $H \subset K$, $|H| = 2$. Let $e_1, e_2, e_3 \in K$. Then G is a figure-of-eight multigraph and can be edge-coloured with three colours with e_1, e_2 and e_3 receiving different colours if and only if $H \not\subset \{e_1, e_2, e_3\}$ and $\{e_1, e_2, e_3\} \not\subset K \setminus H$.

We have a result similar to Theorem 4 in the case when $|H| = 4$.

Theorem 7. Let G be a skew cubic bipartite multigraph with skew cutset K. Let $H \subset K$ be a cutset and let $|H| = 4$. Then G is a figure-of-eight multigraph (and so contains a cutset $K \setminus H$ consisting of two edges).

Proof. This is very similar to Theorem 4 (but this time $\beta + \gamma = 4$).

We now describe the exceptional multigraphs of Theorem 7. Call a skew cubic bipartite multigraph with skew cutset K whose removal separates G into two disjoint submultigraphs G_1 and G_2, a *beermug multigraph* if there is a cutset H such that $|H| = 4$ and $|H \cap K| = 3$ which separates G into two disjoint submultigraphs, whose vertex sets are, say, C and D, with the following properties: Let $\{e\} = H \setminus K$. $\{e\}$ is a cutset of one of G_1 and G_2, say G_1, and the removal of e from G_1 leaves two disjoint submultigraphs whose vertex sets are $V(G_1) \cap C$ and $V(G_1) \cap D$; G_2 is the union of two disjoint submultigraphs whose vertex sets are $V(G_2) \cap C$ and $V(G_2) \cap D$. For some bipartition (A, B) of G, two edges of K join $B \cap C \cap V(G_1)$ to $A \cap C \cap V(G_2)$, two edges of K join $B \cap C \cap V(G_1)$ to $A \cap D \cap V(G_2)$, one edge of K joins $B \cap D \cap V(G_1)$ to $A \cap C \cap V(G_2)$, one edge of K joins $B \cap D \cap V(G_1)$ to $A \cap D \cap V(G_2)$ and e joins $A \cap C \cap V(G_1)$ to $B \cap D \cap V(G_1)$. A beermug multigraph is illustrated in Figure 6.

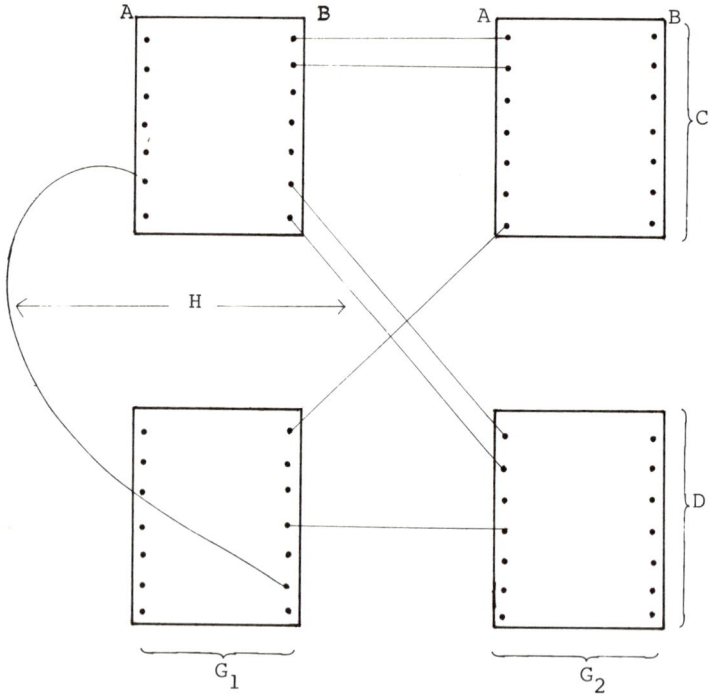

Figure 6. A beermug multigraph.

Theorem 8. Let G be a skew cubic bipartite multigraph with skew cutset K. Let G contain a minimal cutset H with $|H| = 4$ and $|H \setminus K| = 1$. Then G is a beermug multigraph.

Proof. Suppose that H separates G into two disjoint submultigraphs whose vertex sets are C and D. Let e be the edge of $H \setminus K$. We may suppose without loss of generality that the edges of K join vertices of $V(G_1) \cap B$ to vertices of $V(G_2) \cap A$, that $e \in G$, and that e joins a vertex of $B \cap D \cap V(G_1)$ to a vertex of $A \cap C \cap V(G_2)$. Then, since H is minimal, a simple counting argument shows that the remaining edges of H consist of two edges joining $B \cap C \cap V(G_1)$ to $A \cap D \cap V(G_2)$ and one edge joining $B \cap D \cap V(G_1)$ to $A \cap C \cap V(G_2)$. A further simple counting argument then shows that there are 4, 2, 3 and 3 edges of K respectively incident with vertices of $B \cap C \cap V(G_1)$, $B \cap D \cap V(G_1)$, $A \cap C \cap V(G_2)$ and $A \cap D \cap V(G_2)$. Since H is a cutset it then follows that G is a beermug multigraph.

Corollary 9. Let G be a skew cubic bipartite multigraph with skew cutset K. Let K contain a cutset J of G with $|J| = 3$, but no cutset of two edges. Let e_1, e_2, e_3 be three edges of K. Then K can be edge-coloured with three colours with e_1, e_2 and e_3 receiving different colours if and only if G is not a beermug multigraph with e_1, e_2, e_3 being three edges of a cutset H of G with $|H| = 4$ and $|H \setminus K| = 1$.

Proof.

Necessity. If G is a beermug multigraph there are two possibilities for H, and clearly in neither case can G be edge-coloured with three colours so that the edges of $H \cap K$ receive different colours.

Sufficiency. Suppose there is no such edge-colouring with three colours with e_1, e_2, e_3 receiving different colours. Since K contains a cutset J with $|J| = 3$, an easy counting argument shows that at least one of G_1 and G_2, say G_2, is disconnected and, in fact, that $J(= J_1)$ and $K \setminus J(= J_2)$ are both cutsets of G with three edges. In any edge-colouring of G with three colours, it is easy to see that the three edges of J_1 receive different colours, as do the three edges of J_2. Therefore, we may assume that $e_1 \in J_1$ and $e_2, e_3 \in J_2$. In any edge-colouring of G with three colours, e_2 and e_3 will receive different colours; from

Häggkvist's lemma and König's theorem, it is easy to see that e_1 can be made to receive a third colour unless either e_1 and one of $\{e_2, e_3\}$ forms a cutset of G, or unless there is a further edge e such that $\{e, e_1\} \cup \{e_2, e_3\}$ forms a cutset of G. The first possibility is excluded by hypothesis, so the second possibility must hold. But e cannot be a member of K, for then, by Theorem 7, G would be a figure-of-eight graph, and so there would be a cutset of G consisting of two edges of K, a possibility excluded by hypothesis. Therefore $e \notin K$, and so, by Theorem 7, G is a beermug multigraph. This proves Corollary 9.

<u>Corollary 10.</u> Let G be a skew cubic bipartite multigraph with skew cutset K. Let e_1, e_2, e_3 be three edges of K. If a proper subset of K is a cutset of G then either

      (i) G is a figure-of-eight multigraph

or   (ii) G is a beermug multigraph

or   (iii) G is edge-colourable with three colours with e_1, e_2, e_3 receiving different colours.

<u>Proof.</u> An easy counting argument shows that no subset of K with either one or five edges can be a minimal cutset of G. Corollary 10 now follows from Theorems 4 and 6 and from Corollary 9.

We next describe the last of the special skew cubic bipartite multigraphs we shall need to consider. Call a skew cubic bipartite multigraph with skew cutset K, whose removal separates G into two disjoint subgraphs G_1 and G_2, a *cat's-cradle-multigraph* if it contains a minimal cutset M with $|M| = 4$ and $M \cap K = \phi$ which separates G into two disjoint submultigraphs whose vertex sets are P and Q and which has the following properties: The submultigraph induced by P contains two edges of K and G*, the submultigraph induced by Q contains four edges of K; G* has a cutset J with $|J| = 2$ and $J \subset K$ whose removal separates G* into two disjoint submultigraphs whose vertex sets are C and D; G_1 contains a cutset of two edges e_{1C} and e_{1D} which join vertices of $V(G_1) \cap A \cap P$ to vertices of $V(G_1) \cap B \cap C$ and $V(G_1) \cap B \cap D$ respectively, and there are no further edges between any of $V(G_1) \cap P$, $V(G_1) \cap C$ and $V(G_1) \cap D$; similarly G_2 contains a cutset of two edges, e_{2C} and e_{2D}, which join vertices of $V(G_2) \cap B \cap P$ to vertices of $V(G_2) \cap A \cap C$ and $V(G_2) \cap A \cap D$ respectively, and there are no further edges between any of $V(G_2) \cap P$, $V(G_2) \cap C$ and $V(G_2) \cap D$; two edges of K join vertices of $V(G_1) \cap B \cap P$ to vertices of $V(G_2) \cap A \cap P$; finally one edge of K joins each of $V(G_1) \cap B \cap C$ and $V(G_1) \cap B \cap D$ to each of $V(G_2) \cap A \cap C$ and $V(G_2) \cap A \cap D$.

A cat's cradle multigraph is illustrated in Figure 7. Observe that a cat's cradle multigraph also contains four minimal cutsets L such that $|L| = 4$, $|L \cap M| = 2$, $|L \cap K| = 2$, and also observe that both J and $(K \cap E(G^*))\backslash J$ are cutsets of G^* consisting of two edges. In the discussion we shall refer to these as J_1 and J_2.

<u>Theorem 11</u>. Let G be a skew cubic bipartite multigraph with skew cutset K, no proper subset of which is a cutset, and let H be a subset of G with $|H| = 4$ and $H \cap K = \phi$ which separates G into two disjoint submultigraphs Γ_1 and Γ_2, each of which contains three edges of K. Let e_1, e_2, e_3 be three edges of K. Then G can be edge-coloured with three colours with e_1, e_2 and e_3 receiving different colours unless one of Γ_1 and Γ_2 contains precisely one of $\{e_1, e_2, e_3\}$ and G is the cat's cradle multigraph with the submultigraph induced by P containing exactly one of e_1, e_2, e_3 and either $J_1 \subset \{e_1, e_2, e_3\}$ or $J_2 \subset \{e_1, e_2, e_3\}$. [Here P, J_1, J_2 refer to the definition of the cat's cradle multigraph].

<u>Proof</u>. Let $V(\Gamma_1) = E$, $V(\Gamma_2) = F$. A simple counting argument shows that $|H \cap E(G_1)| = 2$ and $|H \cap E(G_2)| = 2$, and that there is one edge of H joining each of the pairs of vertices $V(G_1) \cap A \cap E$ and $V(G_1) \cap B \cap F$, $V(G_1) \cap A \cap F$ and $V(G_1) \cap B \cap E$, $V(G_2) \cap A \cap E$ and $V(G_2) \cap B \cap F$, and $V(G_2) \cap B \cap E$ and $V(G_2) \cap A \cap F$. Let the vertex of $V(G_1) \cap A \cap E$ which is joined to a vertex of $V(G_1) \cap B \cap F$ by an edge of H be denoted by v_{1AE}, and let the other vertices on edges of H be denoted in an obvious similar way.

It is convenient to construct from Γ_1 and Γ_2 two sets $\{\Gamma_1^*, \Gamma_2^*\}$ and $\{\Gamma_1^{**}, \Gamma_2^{**}\}$ of multigraphs as follows. Let Γ_1^* be obtained from Γ_1 by adjoining the two edges $v_{1AE} v_{2BE}$ and $v_{1BE} v_{2AE}$ and let Γ_2^* be obtained from Γ_2 by adjoining $v_{1AF} v_{2BF}$ and $v_{1BF} v_{2AF}$. Let Γ_1^{**} be obtained from Γ_1 by adjoining the two edges $v_{1AE} v_{1BE}$ and $v_{2AE} v_{2BE}$ and let Γ_2^{**} be obtained from Γ_2 by adjoining $v_{1AF} v_{1BF}$ and $v_{2AF} v_{2BF}$.

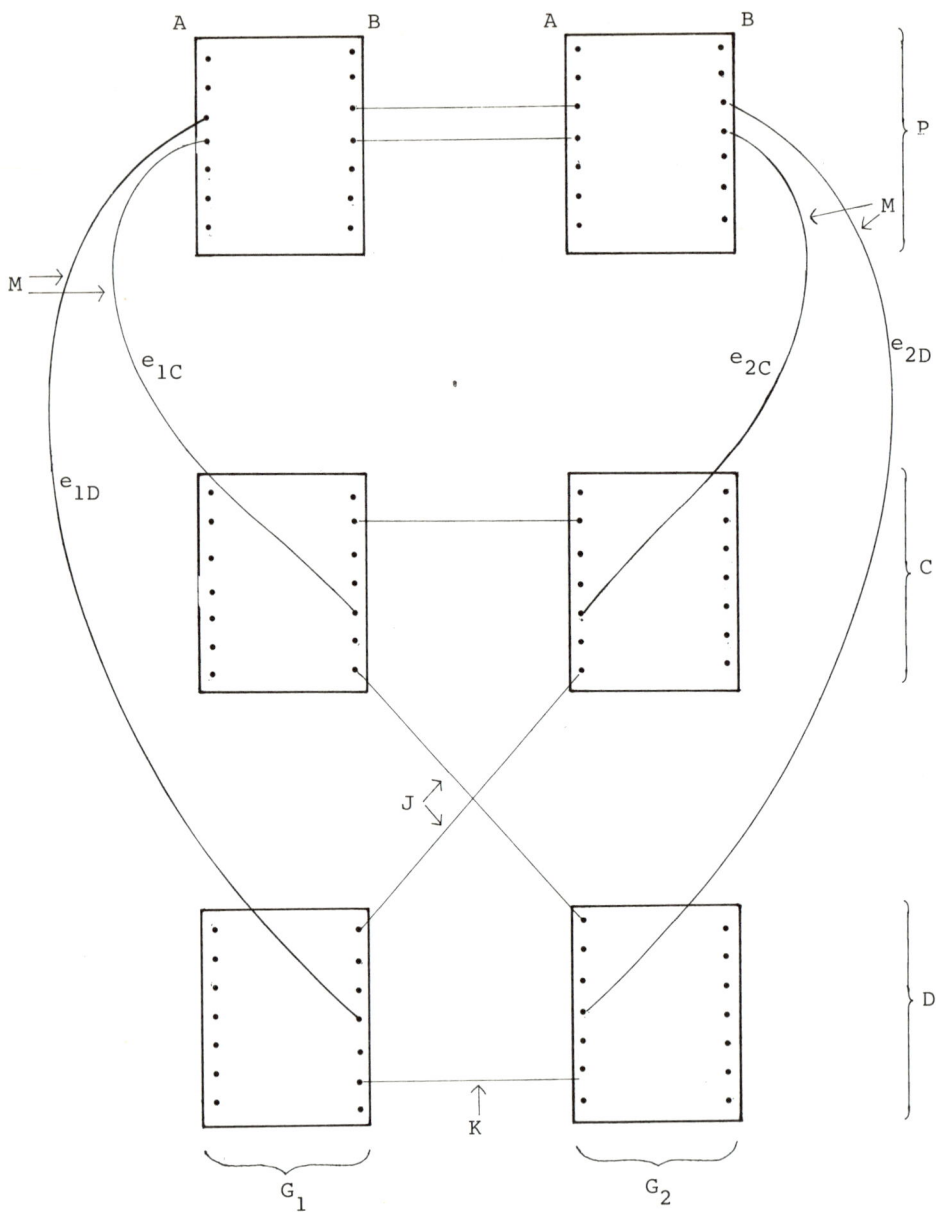

Figure 7. A cat's cradle multigraph.

Suppose $e_1, e_2, e_3 \in E(\Gamma_1)$. A simple counting argument shows that it is not possible for Γ_1^* to contain a cutset of three edges which includes both $v_{1AE} v_{2BE}$ and $v_{1BE} v_{2AE}$. Similarly Γ_2^* cannot contain a cutset of three edges which contains both $v_{1AF} v_{2BF}$ and $v_{1BF} v_{2AF}$. Therefore, by Theorem 1, we may colour the edges of $\Gamma_1^* \cup \Gamma_2^*$ with three colours giving all four adjoined edges the same colour, say α. Then e_1, e_2, e_3 will each receive different colours. We can now reconstruct G with all edges receiving the colours they had in $\Gamma_1^* \cup \Gamma_2^*$ and with the four edges of H each being coloured α. Similarly if $e_1, e_2, e_3 \in E(\Gamma_2)$.

From now we may suppose that both Γ_1 and Γ_2 contain edges of $\{e_1, e_2, e_3\}$. Without loss of generality suppose that $e_1, e_2 \in E(\Gamma_1)$ and $e_3 \in E(\Gamma_2)$. Let f_3 be the third edge of K in Γ_1 and let f_1, f_2 be the other two edges of K in Γ_2.

Suppose that there is no minimal cutset of three edges of Γ_1^* containing f_3 and $v_{1BE} v_{2AE}$ and similarly that there is no minimal cutset of Γ_2^* of three edges containing e_3 and $v_{1BF} v_{2AF}$. Then, by Theorem 1, we may colour $\Gamma_1^* \cup \Gamma_2^*$ so that the four edges $f_3, v_{1BE} v_{2AE}, e_3, v_{1BF} v_{2AF}$ all receive the same colour, say α. Then the other two adjoined edges also receive the colour α and the edges e_1, e_2 receive different colours, not α. As before we may reconstruct G with all edges receiving the colours they had in $\Gamma_1^* \cup \Gamma_2^*$, with all four edges of H being coloured α.

Similarly, for $i = 1,2$, suppose there is no minimal cutset of three edges of Γ_1^* containing $v_{1BE} v_{2AE}$ and e_i and likewise, for $j = 1,2$, there is no minimal cutset of Γ_2^* of three edges containing $v_{1BF} v_{2AF}$ and f_j. Then, by Theorem 1, we may colour $\Gamma_1^* \cup \Gamma_2^*$ so that $v_{1BE} v_{2AE}, v_{1BF} v_{2AF}, e_i$ and f_j receive the same colour, say α. Then the other two adjoined edges also receive the colour α and we can arrange that e_1, e_2 and e_3 all receive different colours. Then, as before, we may reconstruct G with all edges receiving the colours they had in $\Gamma_1^* \cup \Gamma_2^*$, with all four edges of H being coloured α.

Now suppose there is a minimal cutset of Γ_2^* of three edges containing e_3 and $v_{1BF} v_{2AF}$, and also, for each of $j = 1,2$, there is a minimal cutset of Γ_2^* of three edges containing $v_{1BF} v_{2AF}$ and f_j. Then it is easy to see that Γ_2^{**} has a cutset consisting of two edges of K, say k_1 and k_2, such that the removal of $\{k_1, k_2\}$ leaves two disjoint submultigraphs M_1 and M_2 with the properties that k_1 joins a vertex of $V(M_1) \cap B$ to a vertex of $V(M_2) \cap A$, k_2 joins a vertex of $V(M_1) \cap A$ to a vertex of $V(M_2) \cap B$, $v_{1BF} v_{2AF}$ and the third edge k_3 of $K \cap E(\Gamma_2^*)$ join vertices of $V(M_1) \cap B$ to vertices of $V(M_2) \cap A$ and $v_{1AF} v_{2BF}$ joins a vertex of $V(M_2) \cap A$ to a vertex of $V(M_2) \cap B$. This and the corresponding Γ_2^{**} are illustrated in Figure 8.

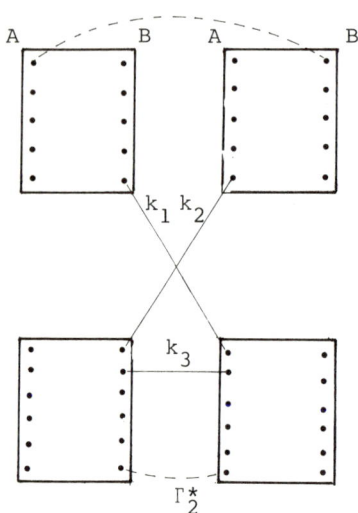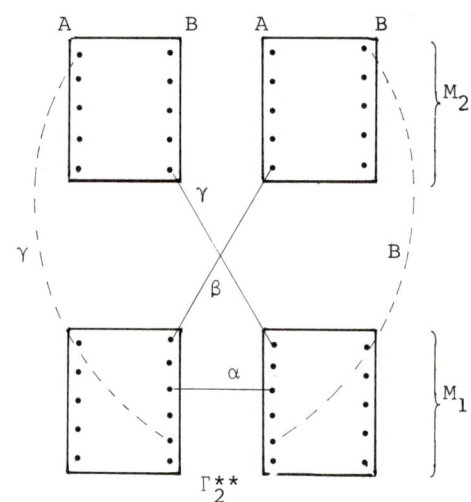

Figure 8.

Either

(A) $\{k_1, k_2\} = \{f_1, f_2\}$, $k_3 = e_3$. [We may suppose that $k_1 = f_1$, $k_2 = f_2$],

or

(B) $\{k_1, k_2\} = \{e_3, f_i\}$, $k_3 = f_j$, where $\{i, j\} = \{1, 2\}$. [We may suppose that $k_1 = e_3$, $k_2 = f_1$, $k_3 = f_2$].

It is easy to check that Γ_2^* cannot be edge-coloured with three colours with the two adjoined edges receiving the same colours, and that the only way of edge-colouring Γ_2^{**} (aside from permutation of the colours) with three colours is shown in Figure 8. We shall suppose that Γ_2^{**} is coloured as in Figure 8.

If $e_3 = k_3$ we may edge-colour Γ_1^{**} so that the adjoined edges $v_{1AE} v_{1BE}$ and $v_{2BE} v_{2AE}$ are coloured γ and β respectively and the edges e_1 and e_2 are coloured β and γ in some order (so that f_3 is coloured α) and thence obtain the required edge-colouring of G, except, by Häggkvist's lemma and Theorems 2 and 3, if either Γ_1^{**} contains a cutset of two edges consisting of one adjoined edge and one edge of K or Γ_1^{**} is a double-circle multigraph with $v_{1AE} v_{1BE} \in E_1$, $v_{2BE} v_{2AE} \in E_2$ and $f \in F \cap E(H_1)$ [E_1, E_2, F and H_1 refer to the description earlier of the double-circle multigraph]. The first case is not possible, for it implies that G has a cutset consisting of three edges of K. In the second case G is a cat's cradle multigraph and G cannot be edge-coloured in the required way.

If $k_1 = e_3$ we may colour Γ_1^{**} so that the adjoined edges $v_{1AE} v_{1BE}$ and $v_{2BE} v_{2AE}$ are coloured γ and β respectively and the edges e_1 and e_2 are coloured α and β in some order (so that f_3 is coloured γ) and thence obtain the required edge-colouring of G unless, by Häggkvist's lemma, $\{f_3, v_{2BE} v_{2AE}\}$ is a cutset of Γ_1^{**}. But this case does not arise, for it implies that G has a cutset consisting of three edges of K.

Now consider the case when there is a minimal cutset of Γ_1^* consisting of three edges which contains f_3 and $v_{1BE} v_{2AE}$ and also, for each of $i = 1,2$, there is a minimal cutset of Γ_1^* consisting of three edges which contains $v_{1BE} v_{2AE}$ and e_1. If we notice that, in an edge-colouring of G with three colours, to require that e_1, e_2, e_3 receive different colours is equivalent to requiring that f_1, f_2, f_3 receive different colours, then we see that this case is essentially a duplicate of the previous case. Thus we do not need to consider it separately.

We are left with two cases to consider. The first case is when Γ_2^* contains: a minimal cutset of three edges that includes $v_{1BF} v_{2AF}$ and f_1; a minimal cutset of three edges that includes $v_{1BF} v_{2AF}$ and f_2; and no minimal cutset of three edges that includes $v_{1BF} v_{2AF}$ and e_3; and simultaneously, Γ_1^* contains: a minimal cutset of three edges that includes $v_{1BE} v_{2AE}$ and f_3; and at least one of e_1 and e_2 (say e_1) does not occur in a minimal cutset of three edges with $v_{1BE} v_{2AE}$. Under these conditions, and recalling that no proper subset of K is a cutset of G, it can easily be verified that either e_3 and f_1 or e_3 and f_2 (say e_3 and f_1) do not occur together in a minimal cutset of three edges in Γ_2^*, and with the exception of the multigraph illustrated in Figure 9 or the analogue of it in which G_1 and G_2 are interchanged [we may suppose that if there is an exception, then it is the multigraph of Figure 9], either e_1 and f_3 or e_2 and f_3 (say e_1 and f_3) do not occur together in a minimal cutset of three edges in Γ_1^*. Then, providing Γ_1^* is not the multigraph shown in Figure 9, using Theorem 1 we can colour $\Gamma_1^* \cup \Gamma_2^*$ so that e_3 and f_1 receive the same colour, say α, and e_1 and f_3 receive the same colour, say β. A simple counting argument reveals that the edges $v_{1AF} v_{2BF}$ and $v_{1AE} v_{2BE}$ must be coloured with α and β respectively, and we can arrange that the edges $v_{1BF} v_{2AF}$, $v_{1BE} v_{2AE}$ and e_2 are coloured with β, α and the third colour, say γ, respectively. As before, we may reconstruct G with all edges receiving the colours they had in $\Gamma_1^* \cup \Gamma_2^*$, with two of the edges in H being coloured α, the other two being coloured β.

Now consider the situation where Γ_1^* is the graph illustrated in Figure 9, where we can assume G_1 and G_2 are as shown. It can be easily verified that since f_1 and f_2 but not e_3 occur in minimal cutsets of three edges that contain $v_{1BF} v_{2AF}$, then either Γ_2^* contains a cutset of two edges consisting of $v_{1AF} v_{2BF}$ and e_3, or Γ_2^* is the graph illustrated in Figure 10. In the latter case, we can immediately reconstruct G. We may then verify that G can be coloured in the required way, remembering that no proper subset of K is a cutset of G. In the former case, removal of $v_{1AF} v_{2BF}$ and e_3 separates Γ_2^* into two disjoint submultigraphs, one of which contains no edges joining G_1 to G_2. This submultigraph must be contained within G_2 for otherwise K would contain a cutset of less than six edges. Now consider Γ_2^{**}. By Häggkvist's lemma we can edge-colour Γ_2^{**} with three colours so that e_3 and $v_{1AF} v_{2BF}$ receive different colours unless they form a cutset of two edges. This exception is impossible, for then e_3, e_2 and f_3 form a cutset in G. Clearly e_3 and $v_{2AF} v_{2BF}$ receive the same colour since they form a cutset of two edges. Now consider Γ_1^{**}. By Theorem 1 we can edge-colour Γ_1^{**} with three colours so that f_3 and $v_{2AE} v_{2BE}$ receive the same colour unless they occur together in a minimal cutset of three edges. This exception is also impossible as G would then contain a cutset consisting of f_1, f_2 and e_2. Now we can arrange for $\Gamma_1^{**} \cup \Gamma_2^{**}$ to be edge-coloured with three colours so that e_3, $v_{2AF} v_{2BF}$ and $v_{2AE} v_{2BE}$ receive the same colour, say α, $v_{1AE} v_{1BE}$, e_1 and $v_{1AF} v_{1BF}$ receive the same colour, say β, and e_2 receives the third colour. So, as before, we can reconstruct G so that all edges receive the colours they had in $\Gamma_1^{**} \cup \Gamma_2^{**}$ with two edges of H being coloured

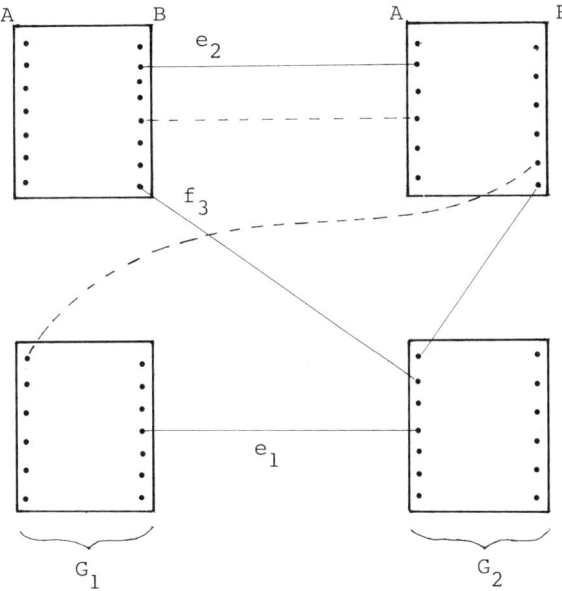

Figure 9.

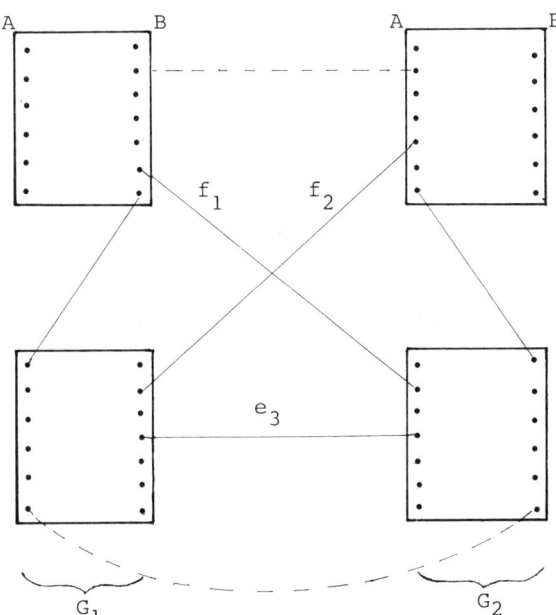

Figure 10.

α, the other two being coloured β

The final case is the same as the case just considered if we replace e_i by f_i and f_i by e_i for $i = 1,2$, and 3. However, this case is essentially a duplicate of the previous case, since requiring e_1, e_2 and e_3 to receive different colours is equivalent to requiring f_1, f_2 and f_3 to receive different colours in an edge-colouring of G. This proves Theorem 11.

3. A CONJECTURE.

We conjecture that Theorem 11 holds without the hypothesis concerning the set H. More specifically:

Conjecture. Let G be a skew bipartite multigraph with skew cutset K, no proper subset of which is a cutset, and let G not be a cat's cradle multigraph. Let e_1, e_2, e_3 be three edges of K. Then G can be edge-coloured with three colours with e_1, e_2 and e_3 receiving different colours.

REFERENCES.

[1] L.D. Andersen, Latin squares and their generalizations. Ph.D. thesis, Reading, 1979.

[2] L.D. Andersen, A.J.W. Hilton and C.A. Rodger, A solution to the embedding problem for partial idempotent latin squares. To appear.

[3] R. Häggkvist, A solution to the Evans conjecture for latin squares of large size. Combinatorics, Proc. Conf. on Combinatorics. Kezthely (Hungary) 1976. Pub. by János Bolyai Math. Soc. and North Holland Pub. Cp. 1978, 495-513.

[4] A.J.W. Hilton, Embedding an incomplete diagonal latin square in a complete diagonal latin square. J. Combinatorial Theory 15 (1973), 121-128.

[5] D. König, Theorie der endlichen und unendlichen Graphen. Chelsea, New York, 1950.

SOME COLOURING PROBLEMS AND THEIR COMPLEXITY

A.J. MANSFIELD and D.J.A. WELSH

Mathematical Institute
Oxford, England

§1. INTRODUCTION

We consider graph colouring problems at different levels of the complexity hierarchy and use them to illustrate various complexity levels within the languages of graphs. These are the problems:

(1)   Is a graph bipartite?
(2)   Find the chromatic polynomial of a graph.
(3)   Is a graph 3 colourable?

The complexity terminology is fairly standard and follows closely that used by Garey and Johnson [2]; though as will be seen in the next section a certain amount of care needs to be taken when considering space requirements of nondeterministic machines as described in [2].

The first problem, that of testing whether a graph is 2-colourable is introduced principally to clarify the notion of space in nondeterministic machines.

Our second set of problems concern the number of colourings of a graph. We have created distinct recognition (= decision) problems associated with the well known problem of finding the chromatic polynomial of a graph.

Section 5 concerns the relationship between the longstanding complexity conjecture that coNP $\neq$ NP and a graph theorem of Hajós [4]. This section was motivated by conversations with Francois Jaeger based on complexity problems raised in [11]. We no longer believe that this approach will settle the NP = coNP question but it does pose some seemingly intractable graph - theoretic complexity problems.

§2. THE COMPUTATIONAL MODEL

As our model of computation we use a Turing machine M with a read only input tape and a read-write work tape.

If M is a Turing machine which accepts a language over its input alphabet Σ, then for $x \in \Sigma^*$, $t(x)$ denotes the *time* (= number of transitions) taken by M to halt when x is the input. The *space* $s(x)$ taken by M to accept x is the number of squares of the work tape which are used by M in accepting x. If $f(n)$ is any function on the integers we say M *accepts* L in time $f(n)$ (space $f(n)$) if $t(x) \le f(n)$ ($s(x) \le f(n)$) for all inputs x of length n.

L is in the class P = P-TIME if there is a machine M and polynomial p such that M accepts L in time $p(n)$. P-SPACE and LOG-SPACE are defined analogously. It is well known that

$$\text{LOG-SPACE} \subseteq P \subseteq \text{P-SPACE} .$$

A *nondeterministic machine* (NDTM) consists of a Turing machine together with an additional read only *guess tape* with a restriction that each square of the guess tape can only be read once, that is a "*destructive read*" *guess tape*. The machine operates as follows:- for a given input x a *certificate* $c(x)$ (a string from Σ^*) appears (magically?) on the guess tape and henceforth the machine operates as an ordinary deterministic machine using the certificate $c(x)$ to verify that x is a member of the language in question.

The time and space taken by such a nondeterministic machine is the time and space used by its deterministic part.

Example 1. Let L be the language of 3-colourable graphs. Then for a given input graph G, the certificate $c(G)$ could be a 3-colouring of G. This can be checked in time n^2 (where n is the number of vertices of G) and in space $O(n)$.

Example 2. Let $\bar{L}$ be the language of graphs which are not 3-colourable. No one yet knows a certificate for $\bar{L}$ which would result in a nondeterministic polynomial time algorithm to recognise $\bar{L}$.

Define NP to be the class of language recognisable in polynomial time by nondeterministic machines. It is well known that

$$P \subseteq NP \subseteq \text{P-SPACE},$$

but it is conjectured that

$$P \ne NP \ne \text{P-SPACE}$$

Define the class co-NP to be the set of languages L such that the complementary language $(\Sigma^* \setminus L)$ is a member of NP. Again

$$P \subseteq \text{co-NP} \subseteq \text{P-SPACE.}$$

but it is conjectured that

$$NP \neq \text{co-NP.}$$

Examples 1 and 2 highlight this problem.

Similarly we define NLOG-SPACE (NL) to be the class of languages recognisable in log space by a non-deterministic machine. Again

$$L \subseteq NL \subseteq P,$$
$$L \subseteq \text{co-NL} \subseteq P,$$

and it is conjectured that the inclusions are proper, and that $NL \neq \text{co-NL}$.

Cook's fundamental theorem [1] shows the existence of a class of hardest problems in NP called NP-*complete problems*. These have the property that if any one of them could be shown to be a member of P then NP = P. Finally a problem π, not necessarily a language recognition problem, is NP-*hard*, if it is provably as difficult as some NP complete problem.

Hardest (i.e. complete) problems also exist in NL, P and P-SPACE but they do not concern us here.

§3. 2-COLOURABILITY

Testing whether or not a graph is 2-colourable is one of the easiest algorithmic problems in graph theory. Call the language of 2-colourable graphs BIPARTITE (otherwise NONBIPARTITE).

(3.1) BIPARTITE $\in$ P.

Proof. For an n-graph there is an $O(n^2)$ algorithm.

(3.2) NONBIPARTITE $\in$ NL

Proof. Guess an odd circuit on the guess tape, call it $v_1, \ldots, v_m$. The deterministic verification consists of:

(a) Store v_1 on the work tape.
(b) Read v_i check adjacency to v_{i-1}.
(c) Store v_i and the parity of the number of vertices we have read so far on the work tape, and wipe out v_{i-1}.
(d) Check that $v_m = v_1$ and that parity is odd.

<u>Problem</u>. As far as we know there is no nondeterministic logspace algorithm for BIPARTITE. In fact we conjecture

(3.3) $\qquad$ BIPARTITE $\notin$ NL,

<u>Note.</u> If we did not stipulate that the guess tape in our nondeterministic machine could only be read once then it would be easy to show that BIPARTITE is a member of NL. All we would do is guess the 2-colouring and check that for each edge of the given graph its endpoints had different colours. We can certainly do this in logspace if we are allowed to read symbols on the guess tape more than once! Indeed this nondeterministic algorithm would also work for 3-COLOURABILITY. Thus we have:

(3.4) If the guess tape in a NDTM is read <u>nondestructively</u> then such a machine can accept in log space any language in NP.

§4. THE CHROMATIC POLYNOMIAL

Consider the problem of computing the chromatic polynomial of a graph G. Clearly this is an NP-hard problem since it contains the problem of finding the chromatic number of G which is known to be NP-complete.

We can formulate this as the following decision problems.

CHROMPOLY

<u>INPUT</u>: Graph G and integers $a_1, \ldots, a_n$

<u>QUESTION</u>: Is the chromatic polynomial $P(G; \lambda)$ of G given by

$$P(G; \lambda) = \sum_{i=1}^{n} a_i \lambda^i ?$$

NUMCOL

<u>INPUT</u>: Graph G, integers x and y

<u>QUESTION</u>: Is $P(G; x) = y$?

Others, notably Gill [3], Simon [8] and Valiant [10] have considered classes of counting problems similar to the above. As defined here such counting problems can be considered within the hierarchy of L, P, P-SPACE etc.

(4.1) **Proposition.** CHROMPOLY $\in$ NP $\iff$ NUMCOL $\in$ NP

Proof. Let A be a nondeterministic polynomial time algorithm for CHROMPOLY.

Then for a given input (G,x,y) to NUMCOL consider the following nondeterministic algorithm

  (i)   Guess $a_1,\ldots,a_n$
  (ii) Verify (using A) that $P(G;\lambda) = \sum a_i \lambda^i$
  (iii) Verify that $\sum a_i x^i = y$.

This is a polynomial length nondeterministic algorithm for NUMCOL.

Conversely suppose that NUMCOL $\in$ NP.

Consider an instance $<G; a_1,\ldots,a_n>$ of CHROMPOLY.

The following is a nondeterministic algorithm for checking the membership of $<G; a_1,\ldots,a_n>$ in CHROMPOLY.

  (i)   Guess $(b_1,\ldots,b_n)$
  (ii) Verify (in polynomial time) that $P(G;k) = b_k$ for $1 \leq k \leq n$.
  (iii) Solve the linear equations $\sum a_i k^i = b_k$ to check in polynomial time that $<G; a_1,\ldots,a_n>$ is a member of CHROMPOLY.

(4.2) **Corollary.** CHROMPOLY $\in$ NP $\implies$ NP = co-NP.

Proof. The complement of the NP complete problem 3-COLOURABILITY consists of those G for which $P(G;3) = 0$ which is a special case of NUMCOL.

This is evidence that CHROMPOLY $\notin$ NP. It seems equally unlikely that CHROMPOLY $\in$ co-NP but we have no formal evidence for this, analogous to the above corollary.

Indeed we make the further conjectures about the status of CHROMPOLY (and NUMCOL)

(4.3) It does not belong to any level of the Meyer-Stockmeyer hierarchy (see [2])

(4.4) It is not P-SPACE complete, (though it obviously is a member of P-SPACE).

We remark that if CHROMPOLY is a member of any level of this hierarchy then the whole class of decision problems based on the class #P (studied by Valiant [10]) is contained in this same level.

§5. THE CURIOSITY OF HAJOS' THEOREM

Consider the theorem of Hajós [4] which gives necessary and sufficient conditions for a graph to be k-colourable.

In the notation of Ore [6] if G_1 and G_2 are two disjoint graphs and

$$e_1 = (a_1, b_1), \quad e_2 = (a_2, b_2)$$

are edges in G_1 and G_2 respectively we construct the *conjunction* (or *series connection* or *Hajós join*) of G_1 and G_2 by identifying the vertices a_1 and a_2, to a single vertex, by deleting the edges e_1 and e_2 and introducing a new edge (b_1, b_2).

If G_1 and G_2 are disjoint graphs a *merger* of G_1 and G_2 is obtained by first performing a conjunction of G_1 and G_2 and then selecting equicardinal sets $A_1 = \{a_1, \ldots, a_t\} \subseteq V_1$, and $A_2 = \{b_1, \ldots, b_t\} \subseteq V_2$ and identifying the vertex a_i with b_i for each i.

(5.1) <u>Theorem</u>. (Hajós) If G is not (k-1)-colourable then there exists a subgraph of G which is obtainable from copies of K_k by a finite sequence of mergers.

Using this theorem, a graph G can be shown to be not (k-1)-colourable by the following constructive 'proof'. Construct a sequence of graphs G_i ($1 \leq i \leq m$) where $G_1 = K_k$ and each G_i ($2 \leq i \leq m$) is obtainable from some G_j, G_k ($j,k < i$) by a merger and where G_m is a subgraph of G. We will call any such sequence a *proof sequence* for G, and its *length* the integer m.

By Hajós' theorem such a proof sequence always exists (and is finite). Define $h_k(G)$ to be the minimum length of a proof sequence for G when G is not k-1 colourable and to be zero otherwise.

An alternative way of looking at this proof technique is to regard such a proof of the non-colourability of G in k-1 colours as a rooted directed binary tree with root a subgraph H of G which is not (k-1)-colourable, with leaves copies of K_k and with each vertex, a non-(k-1)-colourable graph G_i obtained by a merger of the two graphs G_j, G_k which directly precede G_i in the tree. We call such a tree a *proof tree* for G and for each integer k define $t_k(G)$ to be the minimum number of vertices in such a proof tree where, by convention, we define $t_k(G)$ to be 0 when G is (k-1) colourable.

Clearly the complexities of colourability as measured by either a 'tree proof' of a 'sequence proof' are closely related in pretty much the same way as formula size and circuit size are related in standard complexity (see for example Savage [7]).

The following remarks about proof trees and proof sequences are easily verified.

(5.2) In any minimal proof sequence $\langle G_1,\ldots,G_m \rangle$ the graphs G_i are non-isomorphic.

(5.3) Any proof tree using a tree of m vertices gives a proof sequence of length $\leq m$.

(5.4) Any proof sequence of length m gives a tree proof of size $\leq 2^m - 1$ and this upper bound is best possible.

Example. Consider the sequence

$$G_1 = K_{k+1},$$
$$G_2 = G_1 \nabla G_1,$$
$$G_3 = G_2 \nabla G_2,$$

where ∇ represents the operation of a merger.

This sequence forms a minimal proof that G_3 is not k-colourable but the equivalent tree proof needs a tree with 7 vertices. This idea can be generalised to arbitrary m.

A corollary of (5.3) and (5.4) is

(5.5) $\qquad h_k(G) \leq t_k(G) \leq 2^{h_k(G)}$.

When k = 3, we are looking for proofs of non 2-colourability. Such a proof will basically consist of constructing an odd circuit which is a subgraph of G. Hence we have the easy result:

(5.6) If the smallest odd circuit in G has cardinality 2c+1 then

$$h_3(G) \leq \lceil \log(2c-1) \rceil + 1,$$

$$t_3(G) \leq 2c-1.$$

For $k \geq 4$ however we have no polynomial upper bound.

This probably reflects the fact that whereas 2-colourability can be decided in polynomial time 3-colourability is NP-hard.

Although we have so far been concerned with k-colourability it is a well known fact that the following restricted problem is NP-complete.

INPUT: Planar graph G with no vertex degree exceeding 4.

QUESTION: Is G 3-colourable?

Henceforth therefore we will concern ourselves primarily with the case of 3-colourability and accordingly define the *Hajós number* $h(G)$ of a graph G to be the minimum length of a proof sequence that G is not 3-colourable (if G is not 3 colourable) and to be zero if G is 3-colourable, in other words $h(G) = h_4(G)$.

The following result is almost immediate.

(5.7) Theorem. If NP $\neq$ co-NP then given any polynomial p there exists an infinite sequence $(n_i : 1 \leq i \leq \infty)$ with $n_i < n_{i+1}$ and such that for each i there exists a graph G_i with

(a) $|V(G_i)| \leq n_i$

(b) G_i is planar, with vertex degrees 3 or 4

(c) G_i is edge critically 4-chromatic

(d) the Hajós numbers $h(G_i)$ satisfy $h(G_i) > p(n_i)$.

Based on Hajós' proof of his theorem it is straightforward to get the following upper bound on the Hajós number of a graph.

(5.8) Proposition. For a graph G with less than $n^2/3$ edges,

$$h(G) \leq 2^{n^2/3 - |E(G)| + 1} - 1.$$

Otherwise $h(G) = 1$.

Proof. When $|E(G)| \geq n^2/3$ then by Turán's theorem G must contain K_4 as a subgraph and hence $h(G) = 1$. For fixed n let G_0 be a counterexample with a maximum number of edges. Since G_0

cannot contain K_4 there must exist distinct vertices a,b,c with the edges $e = (a,b)$, $f = (a,c)$ not present in G_0 but with the edge (b,c) in G_0. Consider $G_0 + e$. By assumption

$$h(G_0+e) \leq 2^{n^2/3 - |E(G_0)| - 1}.$$

Let $G_0' + f'$ be an isomorphic copy of G_0 with the edge f' inserted joining the vertices (a′, c′). Conjunct $G_0 + e$ with the graph $G_0' + f'$ in the obvious way (i.e. identifying a with a′) and removing the edges e, f′. Now identifying all the vertices x′, of $G_0' + f'$ with their mates x of $G_0 + e$ we arrive at the graph G_0.

Hence we can form a proof sequence for G_0 consisting of a proof sequence for $G_0 + e$, a proof sequence for $G_0' + f'$ together with a final merger of $G_0 + e$ with $G_0' + f'$.

This gives

$$h(G_0) \leq h(G_0+e) + h(G_0' + f') + 1$$
$$\leq 2^{n^2/3 - |E(G_0)| +1 - 1}$$

which is a contradiction.

Another way of looking at this problem is as follows.

For each positive integer n, define h(n) to be the supremum of h(G) over all n-vertex graphs G. Then by the foregoing:

(5.9) If h(n) is polynomially bounded coNP = NP.

Since it is thought likely that NP ≠ coNP it is unlikely that we should be able to find such a polynomial bound for h(n). What is perhaps more surprising is that we have been unable to find a sequence of graphs which would show that h(n) was not polynomially bounded. Indeed we know very little about h(n).

The following results are obvious.

(5.10) h(1) = h(2) = h(3) = 0

(5.11) h(4) = h(5) = 1,

(5.12) h(6) = 3,

(5.13) h(n) < h(n+1).

while from proposition (5.8) it follows that

(5.14) $h(n) \leq 2^{n^2/3}$

In the terminology of Lovász [5], Hajós' theorem does not give a "good characterisation" of non 3-colourability (see [5] p.58) and accordingly we could expect there to exist a sequence of "bad graphs" whose Hajós numbers were exponentially large. A natural candidate sequence is the sequence of critically 4-chromatic graphs constructed by Toft [9] with $4n$ vertices and at least n^2 edges.

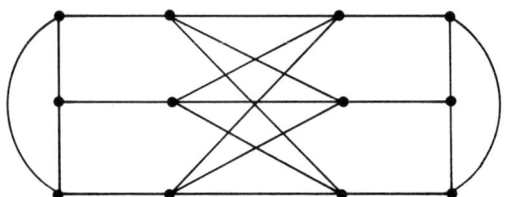

The Toft Graph on 12 vertices

With some difficulty we were able to show that a graph in this sequence with $4n$ vertices has Hajós' number bounded by $A \log n + B$ for suitable constants A and B.

§6. THE COMPLEXITY OF THE HAJÓS NUMBER

Consider the problem of finding the Hajós number of a graph. For example if W_n denotes the wheel with n spokes

(6.1) $h(W_n) = \begin{cases} \lceil \log(n-1) \rceil & n \text{ odd} \\ 0 & n \text{ even.} \end{cases}$

However in general we have found this problem exceptionally difficult apart from a few other special cases.

The difficult is demonstrated for example by the Grötsch graph whose Hajós number we still do not know.

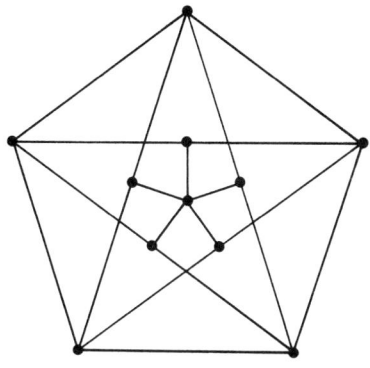

The Grötsch Graph

This is edge critically 4-chromatic and hence has Hajós number $h(G) > 0$. The difficulty we had dealing with this specific graph prompted us to consider the status of the problem:

HAJÓS NUMBER

INPUT: Graph G and positive integer t (in binary form)

QUESTION: Is the Hajós number $h(G) \le t$?

It is obvious that HAJÓS NUMBER is an NP-hard problem for when $t = 0$ it becomes 3-COLOURABILITY. However, the problem does not seem to be in NP, indeed at the moment we have no proof that it is in P-SPACE.

The difficulty in resolving the status of HAJÓS NUMBER is that it does seem to depend heavily on the value of t given in the input. If t is exponential in the order of the graph G then naively it seems there is no way to avoid having to store at least αt ($\alpha > 0$) of the graphs G_i in a Hajós sequence for G. Thus it seems that it may be difficult to settle:

(6.2) <u>Problem</u>. Is HAJÓS NUMBER a member of P-SPACE?

It is however easy to prove:

(6.3) If $h(n)$ is bounded by a polynomial then HAJÓS NUMBER not only belongs to P-SPACE but also to NP.

Acknowledgement.

We would like to acknowledge several stimulating conversations with Francois Jaeger on the subject of Hajós' theorem.

REFERENCES

[1] COOK, S.A., The complexity of theorem proving procedures, *Proc. 3rd Ann. ACM Symp. on Theory of Computing*, Association for Computing Machinery, New York, (1971), 151-158.

[2] GAREY M.R. and JOHNSON D.S., *Computers and Intractibility* W.H. Freeman (San Francisco) (1979).

[3] GILL J.T. III Computational complexity of probabilistic Turing machines, *SIAM J. Comput. (1977)* 6, 675-695.

[4] HAJÓS G. Über eine Konstruktion nicht n-färbarer Graphen. *Wiss. Zeitschr.* Martin Luther Univ. Halle-Wittenberg (1961) A 10, 116-117.

[5] LOVÁSZ L. *Combinatorial Problems and Exercises*, North Holland (1979).

[6] ORE O. *The Four-Color Problem*, Academic Press, New York and London (1967).

[7] SAVAGE J.E. *The Complexity of Computing*, John Wiley & Sons, New York, (1976).

[8] SIMON J. On the difference between one and many, *Springer Lecture Notes in Computer Science* 52 (1977) 480-491.

[9] TOFT B. On the maximal number of edges of critical k-chromatic graphs. *Studia Sci. Math. Hung.* 5 (1970) 461-470.

[10] VALIANT L.G. The complexity of enumeration and reliability problems *SIAM J. Computing* 8 (1979), 410-421.

[11] WELSH D.J.A. Matroids and combinatorial optimisation, *Lectures: Proc. of Varenna Conf.* 1980 (ed. A. Barlotti) C.I.M.E. (to be published).

A HAMILTONIAN GAME

ALEXIS PAPAIOANNOU

University of Cambridge
England

Let P be a property of graphs of order n. We consider the following game, called *Game P*, played on a *board* V consisting of n *vertices*. There are two players: Green and Red, and they move alternately, starting with Green. A *move* of Green is joining by a *green edge* two vertices of V that are non-adjacent at that stage, and a move of Red is joining two vertices by a *red edge*. After $\lceil \frac{1}{2}\binom{n}{2} \rceil$ moves of Green and $\lfloor \frac{1}{2}\binom{n}{2} \rfloor$ moves of Red all pairs of V are edges, some green, some red. Green wins Game P if the graph formed by the $\lceil \frac{1}{2}\binom{n}{2} \rceil$ green edges has property P, otherwise it is a win for Red. [In fact the game may end well before all the vertices of V are joined by some edge, namely Green wins as soon as all graphs of size $\lceil \frac{1}{2}\binom{n}{2} \rceil$ on V that contain the green edges have property P and Red wins when no graph of size $\lceil \frac{1}{2}\binom{n}{2} \rceil$ on V that contains no red edge has property P.] This variant of an Achievement Game of Harary [4] was suggested by Bollobás.

In this note we examine Game P for two properties: that of having a Hamilton path and that of having a Hamilton cycle. The first is the *Hamilton Path Game* and the second is the *Hamilton Cycle Game*.

The Hamilton path game is easily analysed. For $n \leq 3$ *Green wins* trivially. However, for $n = 4$ *Red has a winning strategy* by countering each move of Green with the unique edge independent of the last move of Green. In this way the game ends in one of the two colourings of K^4 shown in Figure 1, neither of which contains a Green path of length 3.

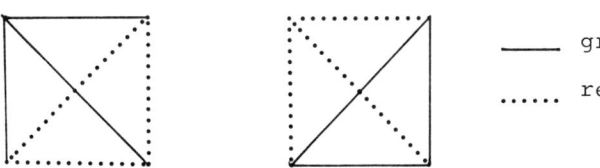

Figure 1. The end of the game on four vertices

Theorem 1. For $n \geq 5$ Green has a winning strategy for the Hamilton Path Game.

Proof. We shall apply induction on n, going from n to $n+2$. The main difficulty is the proof of the theorem for $n = 5$ and 6. We give the details only for $n = 5$.

Suppose $V = \{1,2,3,4,5\}$. We may assume that after two moves of Green and one move of Red, 12 and 45 are the green edges and 23 is the red edge (see Figure 2a). Red can make four essentially different moves: 13, 34, 15 and 25. In each of these cases Green counters with 35 (see Figures 2b, 2c, 2d and 2e). After these moves

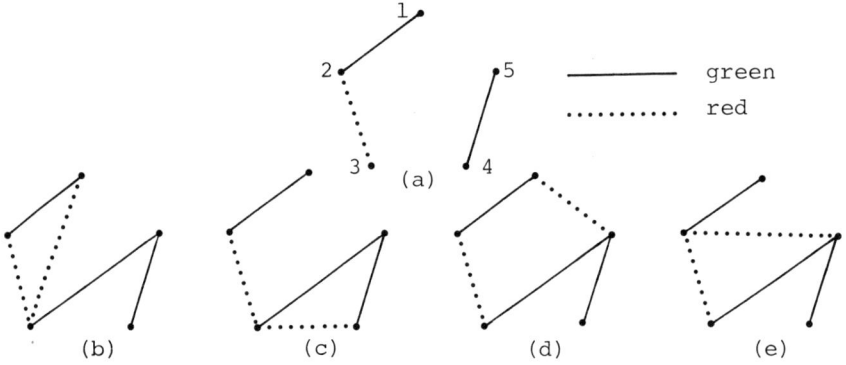

Figure 2. The graphs after five moves

Green has a double threat to join up the paths 354 and 12 to a Hamilton path: 41 and 42. As the next move of Red eliminates at most one of these possibilities, Green can win after his fourth move.

The case $n = 6$ can be dealt with similarly, but there we have to consider ten essentially different sequences of moves.

Suppose now that $n \geq 7$ and the theorem holds for boards with $n-2$ vertices. Green can make sure that after two moves of Green and one move of Red the green edges are $a_1 a_2$ and yz and the red edge is $a_1 x$, where a_1, a_2, x, y and z are distinct vertices. Set $W = V - \{a_1, a_2\}$

From now on Green plays his winning strategy for $n-2$ vertices on the set W, with yz as his starting move. Thus every move of Red that joins two vertices of W is countered by the move of Green appropriate for the Hamilton Path Game played on W. If a move of Red joins a_i to a vertex w in W then Green answers by joining a_j to w, where $\{i,j\} = \{1,2\}$.

What is the situation when all the edges have been drawn? Green has a w_1-w_2 path P of length n-3 that contains all vertices of W. We may assume that the notation is chosen so that $w_1 \neq x$. Then w_1 is joined by a green edge to one of a_1 and a_2, say a_2, so that $a_1 a_2 w_1 P w_2$ is a green path containing all vertices of V. □

The Hamilton Cycle Game is considerably more cumbersome to analyse. Ideally one would like to prove a result similar to Theorem 1. Strangely enough one can prove the induction step from n to n+2 but there are far too many cases to discuss in order to start the induction for n = 8 and 9 , say.

<u>Theorem 2.</u> If Green has a winning strategy for the Hamilton Cycle Game on a board with n_0 vertices then he has a winning strategy for $n_0 + 2$ vertices.

<u>Proof.</u> We know that Green cannot win the game on a board of four vertices since for n = 4 Red wins the Hamilton Path Game. Furthermore, for n = 5 Red can win the Hamilton Cycle Game by making sure that at the end one of the vertices is incident with at least three red edges. Therefore $n_0 \geq 6$.

Suppose $n = n_0 + 2 \geq 8$ and Green has a winning strategy for n-2 vertices. As in the proof of Theorem 1, Green can easily make sure that after two moves of Green and one move of red ab and yz are green edges and ax is red, where a,b,x,y and z are distinct vertices. Set W = V - {a,b} . After this the strategy of Green is as follows.

(i) On the board W Green plays his winning strategy for the Hamilton Cycle Game, that is if Red joins two vertices of W then Green joins two vertices of W as in the Hamilton Cycle Game played on W . If all adges joining vertices of W have been used, he plays an arbitrary move.

(ii) A Red move of the form aw or bw (w ∈ W) is answered by joining a vertex z ∈ W which is adjacent to a or b by a red edge to a or b , whenever possible. Otherwise Green puts in an edge az or bz , z ∈ W . If that is also impossible, Green joins two vertices of W. Given a choice, Green joins z to that one of a and b which has fewer green edges joining it to W.

Green plays this strategy until the very end, provided he can make sure that when only one vertex of W is not joined to a and b then both a and b are incident with green {a,b} - W edges. If this may not be possible then after a move of Red there is only one vertex of W, say u, not incident with any $\{a,b\} - W$ edge and red edges join all other vertices of W to a, say. Then Green deviates from the strategy given in (ii) to play an *exceptional move*: au. Note that when the exceptional move is made the $\{a,b\} - W$ edges are distributed in one of two essentially different ways, shown in Figure 3a, 3b. From here on Green continues to answer

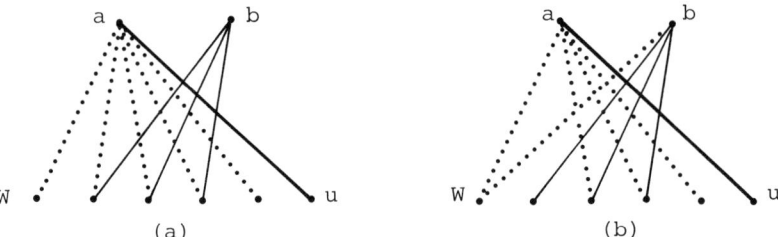

Figure 3. The exceptional move

each $\{a,b\} - W$ edge by one of the same kind, giving preference to an edge incident with a.

What colouring of K^n shall we obtain when all moves are completed?

(A) Suppose Green is not forced to play his exceptional move. Then when the game is over, W has exactly one vertex, say w_1, which is joined to both a and b by a red edge. As Green played his winning strategy on W, there is a green cycle $C = w_1 w_2 \cdots w_{n-2}$ containing all vertices of W. Each w_i, $2 \leq i \leq n-2$, is joined to one of a and b by a green edge and both a and b are adjacent to some w_i by a green edge. Therefore we may assume without loss of generality that for some i, $2 \leq i \leq n-3$, both $w_i a$ and $w_{i+1} b$ are green edges. Then $a w_i w_{i-1} \cdots w_1 w_{n-2} w_{n-3} \cdots w_{i+1} b$ is a green Hamilton cycle, showing that Green has won the game.

(B) Suppose Green has to play his exceptional move. Then the final $\{a,b\} - W$ edges are the edges shown in Figure 3 together with two more red edges and one more green edge. Therefore one of two cases will occur.

(B1) One vertex of W is joined to both a and b by red edges, each other vertex of W is incident with a green $\{a,b\} - W$ edge and both a and b are incident with green $\{a,b\} - W$ edges. This is exactly the same as the distribution described in (A) so Green wins the game.

(B2) There are just two vertices in W, say w_1 and w_2, which are joined to both a and b by red edges, there is just one vertex in W joined to both a and b by green edges, a is incident with 2 green edges joining it to W and b is incident with $(n-2)-3 = n-5 \geq 3$ green edges joining it to W. Furthermore, as Green played his winning strategy on W, there is a green cycle C of length $n-2$ containing all vertices of W.

Suppose $C = w_1 P_1 w_2 P_2 w_1$, that is C is the union of the w_1-w_2 paths P_1 and P_2. We may assume that P_1 has *distinct* internal vertices such that one is joined with a green edge to a and the other is joined with a green edge to b. Since each internal vertex of P_1 is joined to at least one of a and b with a green edge, there are vertices z_1 and z_2 such that they are neighbours on P_1 and both az_1 and bz_2 are green edges. Finally, we may assume that going along P_1 from w_1 to w_2 the vertex z_1 precedes z_2. Then $az_1 P_1 w_1 P_2 w_2 P_1 z_2 b$ is a green Hamilton cycle. □

As we mentioned earlier, at the moment we cannot use Theorem 2 to prove that the Hamilton Cycle Game is a win for Green for all values of n that are not too large. However, we make the following conjecture.

<u>Conjecture</u>. Green has a winning strategy for the Hamilton Cycle Game on a board of at least eight vertices. □

To conclude the paper we prove that the game is a win for Green provided the board is sufficiently large.

<u>Theorem 3</u>. If n is sufficiently large then Green can win the Hamilton Cycle Game on a board of n vertices.

<u>Proof</u>. We need a lemma stating that Green can win a variant of the game if he starts with a certain advantage. The proof is rather cumbersome and will be given elsewhere.

<u>Lemma 4</u>. Suppose the first move of Green is putting in three independent edges and from then on Red and Green alternate as before. Then Green can construct a Hamilton path joining two preassigned vertices. □

Now to prove the theorem assume that $n \geq 600$, say. We shall describe briefly a winning strategy for Green.

(1) The first 200 moves of Green are independent edges. Since

$\binom{150}{2} > 200$, this is possible.

(2) Green uses the next 50 edges to make sure that he has a path P , say from w_1 to w_{50} , of length at most 50 (possibly much less) such that after the first 250 moves of Red every vertex incident with at least 25 red edges is an inner vertex of P . To obtain such a path we make use of the fact that if a green path ends in x and none of wx , wy and wz is a red edge then first adding xw and then one of wy or wz , the green path can be continued to end in y or z . Furthermore, if a green path ends in x , y is a given vertex and u,v are isolated vertices , the first adding ux and then vy , our path can be extended to y by adding either xuvy or xvy .

What do we know about the graph that will arise after the first 250 moves of Green and the first 250 moves of Red? There are at least 200-51 independent green edges having no vertex in common with the green w_1-w_{50} path P. If xy is such a green edge then x and y together are incident with at most 48 red edges. By a theorem of Corrádi and Hajnal [2] (see also [3] and [1;Ch.V]) we find that these independent edges can be partitioned into 49 classes as equal as possible such that no red edge joins endvertices of the edges belonging to the same class. To see this we apply the Corrádi-Hajnal theorem to the graph obtained from the graph of red edges spanned by the independent green edges in which each of our green edges have been contracted to a vertex. In this contracted graph every vertex has degree at most $2 \cdot 24 = 48$, hence the 49 classes. As the classes are as equal as possible, each of them has at least $\lfloor 149/48 \rfloor = 3$ edges.

Let W'_j be the set of vertices of the edges in the ith class, $i = 1,2,\ldots,49$. Note that $|W'_i| \geq 6$. Since w_1 is incident with at most 24 red edges, we may assume that no red edge joins w_1 to W'_1 . Similarly we may assume that no red edge joins w_{50} to W'_{49} . After a suitable renumbering of the sets W'_i we can enlarge W'_i to a set W_i such that

$$V = V(P) \cup \bigcup_{i=1}^{49} W_i ,$$

$$\left(\bigcup_{i=1}^{49} W_i\right) \cap V(P) = \{w_1, w_{50}\}, \quad w_1 \in W_1 , \quad w_{50} \in W_{49} ,$$

$$W_{i-1} \cap W_i = \{w_i\} , \quad 2 \leq i \leq 49 ,$$

$$W_j \cap W_i = \emptyset , \quad 1 \leq j < i \leq 49 .$$

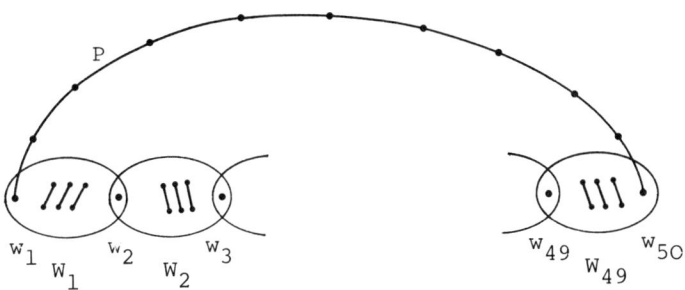

Figure 4. The sets W_i and some of the green edges

To find such sets W_i note that if a vertex w is not joined to either W'_k or W'_ℓ then w can play the role of w_i if W'_k is renumbered W'_{i-1} and W'_ℓ is renumbered W'_i. Furthermore, there are at least $600 - 2 \cdot 250 - 49 = 51$ vertices that are not inner vertices of P and are incident with no red edges. Having found distinct vertices $w_2, w_3, \ldots, w_{49} \in V - V(P)$ such that the sets $W_1^* = \{w_1, w_2\} \cup W_1, W_2^* = \{w_2, w_3\} \cup W_2, \ldots, W_{49}^* = \{w_{49}, w_{50}\} \cup W_{49}$ are independent in the graph spanned by the red edges, one by one we enlarge the W_i^*'s so that they satisfy the remaining conditions. This is easily done since a vertex in $V - V(P)$ is incident with at most 24 red edges.

(3) Having decided on an appropriate choice of the sets $W_1, W_2, \ldots, W_{49}$, Green is concerned only with the red edges joining vertices of the same set W_i. He plays his strategy that on each board W_i guarantees him a green path from w_{i-1} to w_i with vertex set W_i. The union of P and these 49 paths is a Hamilton cycle on the entire board. □

With more care the bound 600 can be lowered but our method is unlikely to provide a bound that is close to being best possible.

REFERENCES

[1] B. Bollobás, *Extremal Graph Theory*, London Math. Soc. Monographs No.11, Academic Press, London and New York, 1978.

[2] K. Corrádi and A. Hajnal, On the maximal number of independent circuits in a graph, Acta Math. Acad. Sci. Hungar. 14 (1963) 423-439.

[3] A. Hajnal and E. Szemerédi, Proof of a conjecture of P.Erdős, in *Combinatorial Theory and its Applications*, vol.II (P.Erdős, A. Rényi, V. Sós, eds.), Colloq. Math. Soc. J. Bolyai $\underline{4}$, North Holland, Amsterdam-London, 1970, 601-623.

[4] F. Harary, Achievement and avoidance games for graphs, in *Graph Theory* (B. Bollobás, ed.), Annals of Discr. Math. 1982.

FINITE RAMSEY THEORY AND STRONGLY REGULAR GRAPHS

J. SHEEHAN

University of Aberdeen
Scotland

We discuss some connections [5], [6], [9], [10] between strongly regular graphs and finite Ramsey theory. All graphs in this paper are both finite and simple. Let G_1 and G_2 be graphs. Then *the Ramsey number*, $r(G_1, G_2)$, *of* G_1 *and* G_2 is the smallest integer n such that in any 2-colouring (E_1, E_2) of the edges of K_n either $<E_1> \supseteq G_1$ or $<E_2> \supseteq G_2$. So, thinking of E_1 and E_2 as being "red" and "blue" edges respectively, if the edges of K_n are coloured red and blue then there exists either a red G_1 or a blue G_2. Furthermore, since n is minimal, there must exist a graph G on n-1 vertices so that $G \not\supseteq G_1$ and its complement $\bar{G} \not\supseteq G_2$. A *strongly regular graph with parameters* (v,k,λ,μ) (or more briefly we say a (v,k,λ,μ)-graph) is a graph which is regular of degree k on v vertices and is such that there exist exactly λ (μ) vertices mutually adjacent to any two distinct adjacent (non-adjacent) vertices. Excellent elementary introductory articles on strongly regular graphs by Cameron and Seidel appear in [1] and [2] respectively. Notice that if G is a (v,k,λ,μ)-graph then $\bar{G}$ is a $(v, v-1-k, v-2k+\mu-2, v-2k+\lambda)$-graph. Now let B_n ($n \geq 1$) denote the graph $K_2 + \bar{K}_n$ (see [8] for notation). Then the interaction between strongly regular graphs and Ramsey theory which we wish to discuss is made formally by the following observation.

OBSERVATION. If there exists a (v,k,λ,μ)-graph G then

$$r(B_{\lambda+1}, B_{v-2k+\mu-1}) \geq v + 1.$$

This follows since $G \not\supseteq B_{\lambda+1}$, $\bar{G} \not\supseteq B_{v-2k+\mu-1}$ and G has exactly v vertices. Now we can consider this inequality from two viewpoints. If a particular (v,k,λ,μ)-graph exists then this determines a lower bound for the corresponding Ramsey number. We give an example of this approach in section 1. On the other hand if we can independently determine an upper bound for a particular Ramsey number this gives some information as to the non-existence of strongly regular graphs with the appropriate parameters (and usually of course to the non-existence of a much larger class of graphs). We take this viewpoint in section 2. In section 3, which might simply be regarded as an appendix, we give a proof of one of the theorems quoted in section 2. We do not recommend the reader to pursue all the details of the proof but simply to note its elementary character. In the final section we make two conjectures.

Section 1

We prove in [9] the following theorem and corollary:-

Theorem 1. If $2(m+n) + 1 > (n-m)^2/3$ then $r(B_m, B_n) \leq 2(m+n+1)$. By refinement, $r(B_{n-1}, B_n) \leq 4n - 1$ and, if $n \equiv 2 \pmod{3}$, $r(B_{n-2}, B_n) \leq 4n - 3$.

Corollary. If $4n + 1$ is a prime power, then $r(B_n, B_n) = 4n + 2$. If $4n + 1$ cannot be expressed as the sum of two integer squares, then $r(B_n, B_n) \leq 4n + 1$. In the first example of the latter, $r(B_5, B_5) = 21$.

We can indicate the main idea of the proof of the theorem by directly proving the Corollary. This proof illustrates the first viewpoint on the observation made in the introduction.

Proof (of the Corollary).

Let $p, n \geq 1$. Suppose $K_p \not\rightarrow (B_n, B_n)$. Let (E_1, E_2) be a 2-colouring of the edges of K_p so that $\langle E_i \rangle \not\supseteq B_n$ ($i = 1, 2$). Let M be the number of monochromatic triangles produced by this colouring. Then a classical result of Goodman [7] gives

$$M \geq (p(p-1)(p-5))/24. \tag{1}$$

On the other hand since on each red (blue) edge there exist at most $n-1$ red (blue) triangles

$$M \leq (|E_1|(n-1) + |E_2|(n-1))/3 = (p(p-1)(n-1))/6. \tag{2}$$

From (1) and (2), $p \leq 4n + 1$. Hence $r(B_n, B_n) \leq 4n + 2$. Suppose $p = 4n + 1$. Then equality holds in (1) and (2). Write $G = \langle E_1 \rangle$. Goodman's result also tells us that since equality holds in (1), G is regular of degree $2n$. Equality in (2) implies that on each edge of G there are exactly $n - 1$ triangles and on each edge of $\bar{G}$

there are exactly n-1 triangles. So G is a (4n+1, 2n, (n-1,n)-graph. Hence, using our observation, $r(B_n,B_n) = 4n + 2$ if and only if there exists a (4n+1), 2n, n-1, n)-graph. Such graphs [4] are called conference graphs and are well known to exist if $4n + 1$ is a prime power. No such graph exists if $4n + 1$ cannot be expressed as the sum of two integer squares. In the first example of the latter $r(B_5,B_5) = 21$. This is proved by giving [9] a direct construction of a graph on 20 vertices with the required properties. □

Section 2

We have proved [6], [9],

Theorem 2.

(i) $r(B_1,B_n) = 2n + 3$ $(n \geq 2)$

(ii) $2n + 3 \leq r(B_2,B_n) \leq \begin{cases} 2n + 6 & (2 \leq n \leq 11) \\ 2n + 5 & (12 \leq n \leq 21) \\ 2n + 4 & (22 \leq n \leq 37) \\ 2n + 3 & (n \geq 38). \end{cases}$ □

Corollary. $r(B_2,B_n) = 2n + 6$, $n = 2,5,11$.   □

Theorem 3. Suppose $1 \leq k \leq n$. Then $r(B_k,B_n) = 2n + 3$ for $n \geq (k-1)(16k^3 + 16k^2 - 24k - 10) + 1$.

Comment. We give a proof of Theorem 3 in the next section. Otherwise all proofs (see [6]) are omitted. We simply interpret these results (or at least some of them) in the context of our viewpoint that upper bounds for Ramsey numbers provide information as to the existence of certain strongly regular graphs.

Let $t \geq 1$. Write $n = 3t - 1$ and let $G(n)$ denote (if it exists) a (6t+3, 2t+2, 1, t+1)-graph. Then if $G(n)$ exists $r(B_2,B_n) \geq 2n + 6$ and so by Theorem 2(ii), $r(B_2,B_n) = 2n + 6$. Now someone (see Cameron [1]) with knowledge of the theory of strongly regular graphs would proceed as follows. If $G(n)$ exists then (using the so-called *integrality condition*)

$$\frac{(v - 1)(\mu - \lambda) - 2k}{\sqrt{(\mu - \lambda)^2 + 3(k - \mu)}}$$

is an integer, where $k = 2t + 2$, $\lambda = 1$, $\mu = t + 1$ and $v = 6t + 3$.

Hence $t = 1,2,4,10$. The line graph $L(K_{3,3})$ of $K_{3,3}$, the complement of $L(K_6)$ and the line graph of the 27 lines on a cubic surface show respectively that $G(2)$, $G(5)$ and $G(11)$ exist. However as yet we have not determined whether $G(29)$ exists. The only other general necessary condition for the existence of a (v,k,λ,μ)-graph are the so-called *Krein conditions* (see Seidel [2]). This states that if

$$r + s = \lambda - \mu \text{ and } rs = \mu - k \quad (r > s)$$

then

(i) $(r + 1)(k + r + 2rs) \leq (k + r)(s + 1)^2$

(ii) $(s + 1)(k + s + 2rs) \leq (k + s)(r + 1)^2$.

In our case $r + s = -10$, $rs = -11$ and so $r = 1$, $s = -11$. We see that these values of r, s and $k = 22$ do not satisfy the second Krein condition. Hence no $G(29)$ exists.

Now suppose we know nothing about the theory of the existence of strongly regular graphs. Then, by Theorem 2(ii), no $G(n)$ exists for $n \geq 12$. So in particular $G(29)$ does not exist. Hence we do not require the Krein conditions to prove the non-existence of $G(29)$. However against this the existence of $G(8)$ is undecided by our Ramsey theory, whereas the integrality test shows that no $G(8)$ exists.

Generally we may interpret Theorems 2 and 3 as follows:-
"if $r(B_m, B_n) \leq N$ then there exists no $(v,k,m-1,n+2k-v+1)$-graph with $v \geq N$".

Of course a much stronger statement is also true viz., for $v \geq N$ there exists no graph G on v vertices such that (i) on each edge of G there are at most $m - 1$ triangles and (ii) on each edge of $\bar{G}$ there are at most $n - 1$ triangles.

Section 3.

Almost all of our notation in this section will be standard ([3], [8]). There is one exception. Let G be a graph and $x \in V(G)$. Then $N(x)$ denotes the neighbourhood of G and for any subset $Y \subseteq V(G)$ we write

$$Y(x) =: N(x) \cap Y.$$

Furthermore if $x_1, x_2 \in V(G)$ we write

$$Y(x_1 \cap x_2) =: Y(x_1) \cap Y(x_2)$$

and

$$Y(x_1 \cup x_2) =: Y(x_1) \cup Y(x_2).$$

This is simply a notational device to restrict the number of symbols used.

In the proof of the theorem below we shall need to consider a 2-colouring (E_1, E_2) of the edges of K_{2n+3} ($n \geq 1$). As above we call the edges of E_1 red and those of E_2 blue. In general a suffix i (i = 1,2) will refer to the i-th colour. For example if $v \in V(G)$, $N_1(v)$ is the red neighbourhood of v and if $Y \subseteq V(K_{2n+3})$, $Y_2(v)$ is the blue neighbourhood of v contained in Y. Again if $Y \subseteq V(K_{2n+3})$, $<Y>_1$ is the subgraph of K_{2n+3} with vertex set Y and edge set consisting of all red edges with both end vertices in Y. We abuse this notation very slightly when we present and use the next lemma but it is only in this context that we shall do this and no confusion should arise.

<u>Lemma</u>. Let $A_1, A_2, \ldots, A_m$ ($m \geq 2$) be subsets of a finite set A. Suppose δ and μ are integers such that for all $i, j \in \{1, 2, \ldots, m\}$, $i \neq j$, $|A_i \cup A_j| \geq \delta$ and $|A_i \cap A_j| \leq \mu$. Then

$$2|A| \geq m\delta - m(m-2)\mu$$

<u>Proof</u>. Choose $i, j \in \{1, 2, \ldots, m\}$, $i \neq j$. Then

$$|A_i| + |A_j| = |A_i \cup A_j| + |A_i \cap A_j|.$$

Summing over all possible such pairs i and j we obtain

$$(m-1)(\sum_{i=1}^{m} |A_i|) = \sum_{i \neq j} |A_i \cup A_j| + \sum_{i \neq j} |A_i \cap A_j|. \quad (3)$$

But

$$|A| \geq |\bigcup_{i=1}^{m} A_i| \geq \sum_{i=1}^{m} |A_i| - \sum_{i \neq j} |A_i \cap A_j|. \quad (4)$$

The lemma now follows from (3), (4) and the definitions of δ and μ. □

<u>Theorem</u>. Suppose k and n are integers such that $1 \leq k \leq n$. Then

$$r(B_k, B_n) = 2n + 3$$

provided $n \geq (k - 1)(16k^3 + 16k^2 - 24k - 10) + 1$.

Proof. We may in fact suppose $k > 1$ since the case $k = 1$ is proved in [9]. Since $K_{n+1,n+1}$ does not contain a B_k and its complement does not contain a B_n we have

$$r(B_k, B_n) \geq 2n + 3. \tag{5}$$

Unfortunately to prove equality is not so straightforward. Suppose $K_{2n+3} \not\to (B_k, B_n)$, $n \geq (k - 1)(16k^3 + 16k^2 - 24k - 10) + 1$. Choose (E_1, E_2) to be a 2-colouring of the edges of K_{2n+3} so that $\langle E_1 \rangle \not\supseteq B_k$ and $\langle E_2 \rangle \not\supseteq B_n$. Choose $\alpha, \beta \in V(K_{2n+3})$ so that $\alpha\beta \in E_2$ and $|N_1(\alpha) \cap N_1(\beta)|$ is as large as possible. Write $D = N_1(\alpha) \cap N_1(\beta)$, $A = N_2(\alpha) \cap N_2(\beta)$, $B = N_2(\alpha) \setminus A$, $C = N_2(\beta) \setminus A$ and $H = B \cup C \cup D$ (see Figure 1, the broken lines indicate red edges). We assume $|B| \geq |C|$.

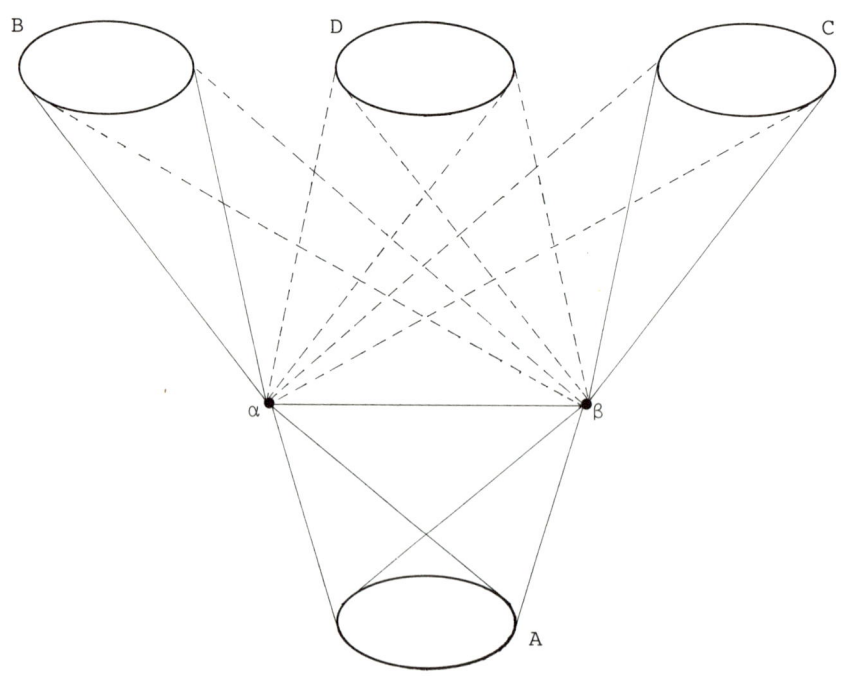

FIGURE 1

We emphasise the choice, in particular the maximality, of $|D|$. It will play a prominent role throughout the subsequent arguments. We proceed by a series of propositions.

Proposition 1.

(1) $|A| \leq n - 1$

(2) $|H| = (2n + 1) - |A| \geq n + 2$

(3) $|B_1(b)| \leq k - 1$ ($b \in B$), $|C_1(c)| \leq k - 1$ ($c \in C$), $|D_1(d)| \leq k - 1$ ($d \in D$)

(4) $|H_1(x)| \leq (k - 1) + |D|$ ($x \in B \cup C$)

(5) $|H_1(d)| \leq 2(k - 1)$ ($d \in D$)

(6) $|H_2(x)| \geq (|H| - 1) - \max\{2(k-1), (k-1) + |D|\}$ ($x \in H$).

Proof. This proof follows directly from the various definitions. For example Proposition 1 (4) (which we abbreviate to P.1.4) is proved by using P.1.3 and the maximality of $|D|$. □

When the reader is in doubt he should refer back to this proposition which will not always be quoted.

Proposition 2. $\langle H \rangle_2 \supseteq K_3$.

Proof. Write $G = \langle B \rangle_2$. Then the minimal degree, $\delta(G)$, of G satisfies, using P.1.3

$$\delta(G) \geq |B| - k. \qquad (6)$$

If $G \not\supseteq K_3$ then, by Turan's theorem [3],

$$|E(G)| \leq |B|^2/4. \qquad (7)$$

Hence, from (6) and (7), $|B| \leq 2k$. Therefore, using P.1.1,

$$|D| = (2n + 1) - (|A| + |B| + |C|)$$
$$\geq (n + 2) - 4k. \qquad (8)$$

We now use the same argument for $K = \: \langle D \rangle_2$. Again if $K \not\supseteq K_3$, $|D| \leq 2k$. So from (8), $n \leq 6k + 2$. This is a contradiction.

Proposition 3. $|D| \geq 2k^2 + 1$.

Proof. Let v_1, v_2, v_3 be the vertices of a triangle in $\langle H \rangle_2$. Write $\theta = \max\{2(k-1), k-1+|D|\}$. Then, from P.1.6, for $i,j \in \{1,2,3\}$, $i \neq j$

$$|H_2(v_i \cap v_j)| \geq |H| - 2(\theta + 1). \tag{9}$$

Since $\langle E_2 \rangle \not\supseteq B_n$, from (9),

$$|A_2(v_i \cap v_j)| \leq (n-1) - (|H| - 2(\theta+1))$$
$$= |A| - n + 2\theta. \tag{10}$$

By the maximality of $|D|$, for all pairs i and j above, $|A_1(v_i \cap v_j)| \leq |D|$. Hence

$$|A_2(v_i \cup v_j)| \geq |A| - |D|. \tag{11}$$

Write $A_i = A_2(v_i)$ ($i = 1,2,3$), $\mu = |A| - n + 2\theta$, $\delta = |A| - |D|$ and $m = 3$. Then from the lemma, (10), (11) and P.1.1

$$3|D| \geq (n+2) - 6\theta. \tag{12}$$

Now suppose $\theta = 2(k-1)$. Then $|D| \leq k-1$ and from (12), $n \leq 15k - 17$, which is not true. On the other hand if $\theta = (k-1) + |D|$ then

$$|D| \geq (n - 6k + 8)/9 \tag{13}$$

and so from (13) and the magnitude of n, $|D| \geq 2k^2 + 1$. □

Proposition 4. $\langle D \rangle_1$ has at least $2k+1$ independent vertices.

Proof. Let $G = \langle D \rangle_1$. Then, from P.1.3 and P.3, G is a graph with maximal degree at most $k-1$ and G has at least $2k^2 + 1$ vertices. The result now follows as an elementary exercise in graph theory. □

Proposition 5. $|D| \geq ((2k-1)|A|)/(2k+1) - (8k^2 - 14k + 3)$.

Proof. Let $v_1, v_2, \ldots, v_{2k+1}$ be distinct independent vertices of $\langle D \rangle_1$. Let $m = 2k+1$ and write $A_i = A_2(v_i)$ ($i = 1, 2, \ldots, m$). Now with minor modifications (allowing for the fact that the v_i's not only belong to H but also to D) we repeat the argument of P.3. Let $i,j \in \{1,2,\ldots,m\}$, $i \neq j$. From P.1.5

$$|H_2(v_i \cap v_j)| \geq 2(|H| - 2 - 2(k-1)) - (|H| - 2).$$

Hence, since $\langle E_2 \rangle \not\subseteq B_n$ and using P.1.2,

$$|A_i \cap A_j| \leq (n - 1) - |H_2(v_i \cap v_j)|$$
$$\leq (n - 1) - |H| + 4(k - 1) + 2$$
$$\leq 4k - 5. \qquad (14)$$

Again, by the maximality of $|D|$ and since $\alpha, \beta \in N_1(v_i) \cap N_1(v_j)$, $|A_1(v_i \cap v_j)| \leq |D| - 2$. Hence

$$|A_i \cup A_j| \geq |A| - |D| + 2. \qquad (15)$$

Write $\delta = |A| - |D| + 2$, $\mu = 4k - 5$. Then the result follows from the lemma using $m = 2k + 1$ and equations (14) and (15). □

Proposition 6. (1) If $x_1, x_2 \in D$ and $x_1 x_2 \in E_2$ then

$$|H_2(x_1 \cap x_2)| \geq n - 4(k - 1)$$

(2) $|A| \geq n - 4(k - 1)$.

Proof. (i) From P.1.5.

$$|H_2(x_1 \cap x_2)| \geq (|H| - 2) - 4(k - 1)$$
$$= (2n - 1) - |A| - 4(k - 1). \qquad (16)$$

The result now follows from P.1.1.

(ii) Since $|D| \geq 2k^2 + 1$ and $\langle D \rangle_1 \not\subseteq B_k$ there exists at least one blue edge $x_1 x_2$ with $x_i \in D$, $i = 1, 2$. The result now follows from (16). □

Proposition 7. $\langle D \rangle_1$ does not contain 2 independent edges.

Proof. Suppose $\langle D \rangle_1$ does contain 2 independent edges. Then we may choose $x_i, y_i \in D$ $(i = 1, 2)$ so that $x_1 y_1$ and $x_2 y_2$ are red edges and such that $x_1 x_2, x_1 y_2, y_1 x_2, y_1 y_2$ are all blue edges. Then, for $i = 1, 2$, since $\alpha, \beta \in N_1(x_i \cap y_i)$ and since $\langle E_1 \rangle \not\subseteq B_k$,

$$|A_1(x_i \cap y_i)| \leq k - 3.$$

Hence

$$|A_2(x_i \cup y_i)| \geq |A| - (k - 3). \qquad (17)$$

Therefore, with no loss of generality, we may suppose that

$$|A_2(x_1)| \geq (|A| - (k - 3))/2. \qquad (18)$$

Now, from (17),

$$|A_2(x_1) \setminus A_2(x_2 \cup y_2)| \leq k - 3. \qquad (19)$$

Therefore from (19), again with no loss of generality, we may suppose that

$$|A_2(x_1 \cap x_2)| \geq (|A_2(x_1)| - (k-3))/2. \qquad (20)$$

Hence, from P.6.1, P.6.2, (18), (20) and since $\langle E_2 \rangle \not\supseteq B_n$

$$n - 1 \geq |A_2(x_1 \cap x_2)| + |H_2(x_1 \cap x_2)|$$
$$\geq (n - 7k + 13)/4 + (n - 4(k-1))$$

i.e., $n \leq 23k - 33$ which is a contradiction. □

<u>Proposition 8</u>. Write $X = B \cup C$ and let $D^* = \{d \in D : |X_1(d)| \geq 1\}$. Then

$$|D^*| \geq |D| - (2k^2 - 6k + 7).$$

<u>Proof</u>. Firstly notice that at most one element of D is isolated in $\langle H \rangle_1$ i.e. $|H_1(d)| \geq 1$ for all but at most one element of D. Otherwise choose two such elements $d_1, d_2 \in D$, $d_1 \neq d_2$. Then, using P.1.2

$$n - 1 \geq |H_2(d_1 \cap d_2)| \geq |H| - 2 \geq n.$$

Now write $G = \langle D \rangle_1$. Then G is a graph with at least $2k^2 + 1$ vertices and maximal degree at most, using P.1.3, $k - 1$. Since G has at most one independent edge xy an elementary exercise in graph theory (see Figure 2) shows that G contains at least $|D| - (2 + (k-2) + 2(k-2)^2)$ isolated vertices.

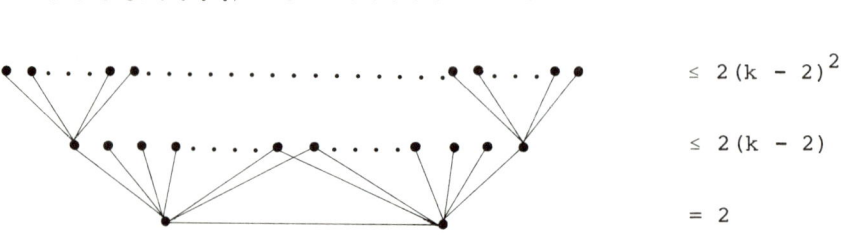

FIGURE 2

This, together with the opening remark, proves the proposition.

Proposition 9. $|D| \leq ((2n + 1 - |A|)(k - 1) + 2k^2 - 6k + 7))/k$.

Proof. Let $E_1(B \cup C, D)$ denote the set of red edges with one end vertex in $B \cup C$ and the other end vertex in D. Then, from P.8,

$$|E_1(B \cup C, D)| \geq |D^*|$$
$$\geq |D| - (2k^2 - 6k + 7). \qquad (21)$$

Since $\langle E_1 \rangle \not\supseteq B_k$ if $x \in B \cup C$, $|D_1(x)| \leq k - 1$. Hence

$$|E_1(B \cup C, D)| \leq |B \cup C|(k - 1)$$
$$= (2n + 1 - |A| - |D|)(k - 1). \qquad (22)$$

The proposition follows from (21) and (22).

Proof of Theorem 3. From Propositions 5 and 9

$$k(2k - 1)|A| - k(2k + 1)(8k^2 - 14k + 3) \leq (2k + 1)(2n + 1 - |A|)(k-1)$$
$$+ (2k + 1)(2k^2 + 6k + 7).$$

Now use P.6.2 to obtain a contradiction to the magnitude of n. Hence $K_{2n+3} \to (B_k, B_n)$ and so, from (5), $r(B_k, B_n) = 2n + 3$. □

A similar [6], but very much more delicate, analysis proves Theorem 2(ii). Theorem 2(i) is proved in [9].

Section 4

We have conjectured in [9].

Conjecture 1. There exists a constant $A > 0$ such that

$$r(B_m, B_n) \leq 2(m + n + 1) + A. \qquad □$$

Our theorems support this conjecture although of course they are a very long way from giving the whole picture. The well-known (253, 112, 36, 60)-graph shows that $r(B_{37}, B_{88}) \geq 254$ and so $A \geq 2$. This conjecture would imply, if true, the truth of:-

Conjecture 2. For any (v, k, λ, μ)-graph there exists a constant A $(A \geq 2)$ such that

$$2(\alpha + \beta) - v \leq A$$

where $\alpha = k - \lambda - 1$ and $\beta = k - \mu$. □

Comment. This conjecture is true for conference graphs. It is also true when $\lambda = \mu$. For readers not familiar with the parameters α and β it is worthwhile recalling in this context that if we write $\ell = v - k - 1$ then since $k(k - \lambda - 1) = \ell\mu$,

$$\frac{\alpha}{\ell} + \frac{\beta}{k} = 1.$$

Finally we would like to mention that if instead of discussing $r(B_m, B_n)$ we consider the Ramsey number $r(K_m + \bar{K}_n)$ then in [5] and [10] conference graphs are used to provide lower bounds. In [10] conference graphs are used to provide lower bounds. In [10] especially the asymptotic lower bounds are discussed in some depth.

REFERENCES

[1] Beineke, L.W. and Wilson, R.J., Selected topics in graph theory, Academic Press, London (1978).

[2] Bollobás, B., Surveys in combinatorics, (Proceedings of the 7th British Combinatorial Conference), C.U.P., Cambridge (1979).

[3] Bondy, J.A. and Murty, U.S.R., Graph theory with applications, Macmillan, London (1976).

[4] Cameron, P.J. and Van Lindt, J.H., Graph theory, coding theory and block designs, C.U.P., Cambridge (1975).

[5] Evans, R.J., Pulham, J. and Sheehan, J., On the number of complete subgraphs contained in certain graphs, J.C.T. (to appear).

[6] Faudree, R.J., Rousseau, C.C. and Sheehan, J. Strongly regular graphs and finite Ramsey theory, (submitted).

[7] Goodman, A.W. On sets of acquaintances and strangers at any party, Amer. Math. Monthly 66 (1959) p.778-793.

[8] Harary, F. Graph theory, Addison-Wesley, Reading, Mass. (1969).

[9] Rousseau, C.C. and Sheehan, J. On Ramsey numbers for books, J.G.T. 2 (1978) p.77-87.

[10] Thomason, A.G. Ph.D. Thesis, Cambridge (1979).

THE CONNECTIVITIES OF A GRAPH AND ITS COMPLEMENT[1]

RALPH TINDELL

Department of Pure and Applied Mathematics
Stevens Institute of Technology
Hoboken, N.J. 07030, U.S.A.

It is an elementary observation that the complement of a disconnected graph must be connected. This fact may be viewed as a necessary condition for a pair $k,\bar{k}$ of nonnegative integers to be realizable as the point-connectivities of a graph and its complement: $k+\bar{k}$ must be positive. This observation leads naturally to the problem of determining the pairs of nonnegative integers $k,\bar{k}$ for which there exists a p-point graph having point-connectivity k and whose complement has point-connectivity $\bar{k}$. In the present paper we show that such pairs are completely determined by the following conditions:
(1) $0 < k + \bar{k} \le p-1$; (2) if $k+\bar{k} = p-1$ then at least one of $\bar{k},k$ is even (3) if one of $k,\bar{k}$ is p-2, then the other is zero.

1. INTRODUCTION

Since the complement $\bar{G}$ of a disconnected graph G is spanned by a complete bipartite graph it must be connected. In the notation of the book [4] by Harary, which we henceforth assume, this may be restated as $\kappa(G) + \kappa(\bar{G}) > 0$. This leads to the question of which pairs of nonnegative integers $k,\bar{k}$ occur as the point-connectivities of a graph and its complement. As we shall see, $k+\bar{k} > 0$ is both necessary and sufficient if the number p of points of the graph is unrestricted. This will be apparent from our solution of the more difficult version of the problem where the number of points is

---

[1] Dedicated to Professor Frank Harary on the occasion of his sixtieth birthday.

specified in advance. Let us say that a triple (p,k,k) is
realizable[1] if there is a p-point graph G with $\kappa(G) + k$ and
$\kappa(\bar{G}) = \bar{k}$; if the graph G also satisfies $\kappa(G) = \delta(G)$ and
$\kappa(\bar{G}) = \delta(\bar{G})$, we will say that the triple is δ-realizable.

The purpose of the present paper is to prove the following characterization of realizable triples. We shall write $(a,b,c) \geq (a',b',c')$ when $a \geq a'$, $b \geq b'$, and $c \geq c'$.

Theorem. An integer triple $(p,k,\bar{k}) \geq (3,0,0)$ is realizable if and only if the following three conditions are satisfied.

(1) $0 < k + \bar{k} < p-1$.
(2) If $k + \bar{k} = p-1$ then one of $k,\bar{k}$ is even.
(3) If one of $k,\bar{k}$ is p-2 then the other is zero.

In section 2 we establish the necessity of conditions (1), (2), and (3) for realizability and show that any p-point graph G with $\kappa(G) + \kappa(\bar{G}) = p-1$ must be regular and have maximum connectivity, which is to say that $\kappa(G) = \delta(G)$, and that the same holds for its complement. In section 3 we state and prove an elegant theorem of Watkins [5] concerning point-transitive graphs[2]. We also introduce an important class of point-symmetric graphs - circulants - and apply Watkin's result to show that specific examples of these graphs have maximum connectivity. These examples are used in section 4 to establish the sufficiency of conditions (1), (2), and (3) for realizability (in fact, for δ-realizability) in the cases where $k + \bar{k} = p-1$. The rest of section 4 is devoted to show how the examples for the extremal case may be modified to yield realizations in the remaining cases.

2. NECESSITY

That the condition $0 < k+\bar{k}$ is necessary for realizability of the triple $(p,k,\bar{k})$ was noted above. Recall that $\delta(\bar{G}) = p-1-\Delta(G)$, where $\Delta(G)$ is the maximum degree of G.

---

[1] In the usage of Boesch and Suffel ([2],[3]), we should perhaps speak of $(p,\kappa,\bar{\kappa})$-realizability, where $\bar{\kappa}$ is a notation for $\kappa(\bar{G})$.

[2] The proof we present is shorter and more accessible than that given in [5].

Thus

$$\kappa(G) + \kappa(\bar{G}) < \delta(\bar{G}) + \delta(\bar{G}) = (p-1) - (\Delta(G) - \delta(G)) \le p-1$$

which establishes the necessity of condition (1). Moreover, we can also note that if $\kappa(G) + \kappa(\bar{G}) = p-1$ then $\Delta(G) = \delta(G)$, so that G is regular and also that $\kappa(G) = \delta(G)$ and $\kappa(\bar{G}) = \delta(\bar{G})$. We gather these observations together as a proposition.

Proposition 1 Let G be a p-point graph with $p \ge 3$ and $\kappa(G) + \kappa(\bar{G}) = p-1$. Then G is regular, $\kappa(G) = \delta(G)$, and $\kappa(\bar{G}) = \delta(\bar{G})$.

From the preceding paragraph we note that if $\kappa(G) + \kappa(\bar{G}) = p-1$ then $\kappa(G)$ and $\kappa(\bar{G})$ cannot be odd since in that case G would be an odd-regular graph on an odd number of points. This establishes the necessity of condition (2). Moreover, when $\kappa(G) + \kappa(\bar{G}) = p-1$, neither $\kappa(G)$ nor $\kappa(\bar{G})$ can be 1 since there is no 1-regular, connected graph on $p \ge 3$ points. This establishes the necessity of (3) and concludes the present section.

3. POINT-SYMMETRIC GRAPHS

A graph G is <u>point-symmetric</u> if for any pair u, v of points of G there is an automorphism ϕ of G with $\phi(u) = v$. We now define and establish the point-symmetry of a class of graphs called circulants. We shall denote by $[x]$ the integer part of the real number x. Given integers $n_1, \ldots, n_r$ with $1 \le n_1 < n_2 \ldots < n_r \le [p/2]$, the p-point circulant with jump sizes $n_1, \ldots, n_r$ is defined to have points $v_0, v_1, \ldots, v_{p-1}$ with v_i and v_j adjacent if and only if one of $j-i, i-j$ is congruent mod p to one of the jump sizes. Note that the above graph is regular, and has degree $2r-1$ if $n_r = p/2$ and has degree $2r$ otherwise. It is simple to show that for any integer m the rotation ρ_m defined by $\rho_m(v_i) = v_{i+m}$, all indices reduced mod p, is an automorphism of any p-point circulant.
Thus circulants are point-symmetric since point v_i may be mapped to point v_j by applying the rotation ρ_{j-i}.

It is somewhat surprising to learn that point-symmetric graphs can fail to have maximum point-connectivity, in that the connectivity can be less than the minimum degree[1]. The circulant on 12 points with

---

[1] One of the earliest examples, a circulant of 15 points, was given by Boesch and Felzer[1].

jump sizes 1,3,4,5 is an example in that it is regular of degree 8 yet has point-connectivity 6. Watkins [5] studied the point-symmetric graphs G with $\kappa(G) < \delta(G)$ and his results provide significant insight into their structure. Moreover his results may be used to establish that certain specific point-transitive graphs do have maximum connectivity. We now present some of Watkins results.

If U is a set of points of a graph G with G-U disconnected, we call U a <u>disconnecting set of G</u>; if the number $|U|$ of points in U equals $\kappa(G)$, then U is a <u>minimum disconnecting set of G</u>. A subgraph B of G is a κ-part of G if B is a component of G-U for some minimum disconnecting set U of G. Note that in this case U must be what we shall call the <u>deleted neighborhood</u> $N^*(B)$, which is defined for arbitrary subgraphs B of G to be the set of points of G which are not in B but are adjacent in G to at least one point of B. The <u>atomic number</u> of G is defined as

$$a(G) = \min \{|B| : \text{is a } \kappa\text{-part of } G\}.$$

Recall that $|B|$ denotes the number of points in B. A κ-part of a graph G having exactly $a(G)$ points is called an <u>atomic part</u> of G. The following is obvious.

<u>Proposition 2</u> For any graph G, $\kappa(G) = \delta(G)$ if and only if $a(G) = 1$.

The next result is a very slight extension of one of Watkins main results.

WATKINS THEOREM: Let A be an atomic part and B a κ-part of a connected graph G. If A∩B is nonempty then A is a subgraph of B.

Proof. The proof is by induction on the number of points in G, and the basis step is trivial. Thus we assume the theorem true for all graphs with fewer then p-points and let B and A be a κ-part and an atomic part, respectively, of a p-point graph G, with A∩B nonempty.
First consider the case where there is a point u in $N^*(A) \cap N^*(B)$. Then W is a disconnecting set of G-u if and only if W∪{u} is a disconnecting set of G. Thus B is a κ-part and A an atomic part of G-u with A∩B nonempty. By our inductive assumption, A is a subgraph of B.

We thus assume that $N^*(A) \cap N^*(B)$ is empty. Now for any two subgraphs A, B of G with nonempty intersection, it is easy to see that $N^*(A \cap B) = (N^*(A) \cap B) \cup (N^*(B) \cap A) \cup (N^*(A) \cap N^*(B))$.

Therefore, $N^*(A \cap B) = S$, where

$$S = (N^*(A) \cap B) \cup (N^*(B) \cap A).$$

We now note that $G-S$ is disconnected since $G-S-(A \cap B)$ contains $G-(N^*(B) \cup B)$, which cannot be empty as B is a κ-part of G. Moreover, we know that $A \cap B$ must contain a component of $G-S$. The very definition of atomic part would force $A \cap B = A$ if we knew that S were a minimum disconnecting set of G, so we may conclude the proof by showing that $|S| = \kappa(G)$.

Since $G-S$ is disconnected we have

(i) $|S| = |N^*(A) \cap B| + |N^*(B) \cap A| \geq \kappa(G)$.

Moreover, since $|N^*(A)| = |N^*(A) \cap B| + |N^*(A)-B| = \kappa(G)$ we have

(ii) $|N^*(B) \cap A| \geq |N^*(A)-B|$.

Now consider the subgraph of $G-N^*(B)-B$ given by

$$H = (A \cup N^*(A)) - (B \cup N^*(B)) = ((A-B)-N^*(B)) \cup (N^*(A)-B).$$

Then

$$|H| = |A-B| - |N^*(B) \cap A| + |N^*(A)-B| \leq |A-B| < |A|$$

by (ii). Since $|G-B-N^*(B)| \geq a(G)$, we see that the subgraph $G-(A \cup N^*(A) \cup B \cup N^*(B))$ is nonempty. It is simple to show that

$$N^*(A \cup B) = (N^*(A)-B) \cup (N^*(B)-A)$$

and hence that $N^*(A \cup B)$ disconnects G. Therefore

$$\kappa(G) \leq |N^*(A)-B| + |N^*(B)-A| \leq |N^*(B) \cap A| + |N^*(B)-A| = |N^*(B)| = \kappa(G).$$

The second inequality in the preceding line follows from (ii). It now follows that $|N^*(B) \cap A| = |N^*(A) - B|$ and hence that $|S| = |N^*(A)| = \kappa(G)$, and the proof is complete.

<u>Corollary 1</u> Distinct atomic parts of a graph G are disjoint.

<u>Corollary 2</u> If A is an atomic part of G and ϕ is an automorphism of G with $\phi(A) \cap A$ nonempty, then $\phi(A) = A$.

Corollary 3 The atomic parts of a point-symmetric graph are themselves point-symmetric graphs.

Corollary 4 Each point of a connected, point-transitive graph G lies in a unique atomic part of G and hence $a(G)$ divides the number of points of G.

We now apply Watkins results to obtain useful structural properties of circulants. Recall that all indices are reduced mod p.

Lemma 1 Let G be a connected circulant on the points $v_0, v_1, \ldots, v_{p-1}$, with $p \geq 4$. Let a be the atomic number of G, let A be the atomic part of G containing v_0, and let $m = p/a$. Then

(i) vertex v_i is in A if and only if m divides i; and

(ii) for each i, $0 < i < a-1$, there is a j with $im < j < (i+1)m$ such that v_j is adjacent to no point of A.

Proof. First we note that when a=1 conclusion (i) is immediate and conclusion (ii) follows from the assumption that G is not complete. Thus assume a >1 so that $m \leq p/2$. Let s be the least positive index with v_s in A. Since the rotation ρ_s maps v_0 to v_s, $\rho_s(A)=A$ by corollary 2 to Watkins Theorem. For each $i \geq 1$ the i-fold application of ρ_s will map v_0 to v_{is} and hence every v_j with j a multiple of s is in A. Moreover if v_b is in A and is <b <(i+1)s, then o < b-is < s and v_{b-is} is in A since $\rho_{is}(v_{b-is})= v_b$. Since this violates our choice of s we may conclude that v_j is in A if and only if s divides j. It then follows that a = p/s and hence that s=m, and conclusion (i) is established.

To prove conclusion (ii) suppose to the contrary that for some i, $0 \leq i \leq a-1$, and every v_j with $im < j < (i+1)m$, v_j is adjacent to some point of A. If v_k is not in A, then there is an integer d, $0 \leq d \leq a-1$, with $dm < k < (d+1)m$. The rotation $\rho_{(i-d)m}$ and maps v_k to v_j, where $j = k + (i-d)m$. Thus $im < j < (i+1)m$ so that v_j is adjacent to some point of A. Since $\rho_{(i-d)m}$ maps A onto itself, v_k is adjacent to some point of A. Thus we have shown that every point of G is either in the atomic part A or in its deleted neighborhood. This contradiction establishes the lemma.

We now use lemma 1 to establish that for certain circulants G, both G and $\bar{G}$ have maximum connectivity.

<u>Proposition 2</u> Let r, p be integers with $p \geq 4$ and $1 \leq r < p/2$. Let G be the p-point circulant with jump sizes $1, 2, \ldots, r$ and, if p is even, let G' be the p-point circulant with jump sizes $1, 2, \ldots, r, p/2$. Then

(i) G and G' both have atomic number 1;

(ii) If $p \geq 4r$ then $\bar{G}$ has atomic number 1; and

(iii) if p is even and $p \geq 4r+3$, then $\bar{G'}$ has atomic number 1.

Proof. Recall that p-point circulants have points $v_0, v_1, \ldots, v_{p-1}$ and indices are always assumed reduced mod p. For any circulant under discussion we shall denote its atomic number by a, the atomic part containing v_0 by A, and set $m=p/a$. Since $a(K_p) = 1$ we may assume that none of the graphs considered is complete—that is, not all possible jump sizes have been used. For each of the graphs $G, G', \bar{G}$, and $\bar{G'}$ our basic method of proof will be to assume $a > 1$ and derive a contradiction to conclusion (ii) of lemma 2, which is to say we will show that for some b, $0 \leq b \leq a-1$, and every j with

$bm < j < (b+1)m$, v_j is adjacent to some point of A.

The proof for case (i) is quite simple. Since G is not complete, $r < [p/2]$ and thus, assuming $a > 1$, v_o must be adjacent to some v_{im} with $0 < im < p/2$. By the definition of G, $im \leq r$ and thus v_o is adjacent to every v_j with $(i-1)m < j < im$, a contradiction. Thus $a(G) = 1$. As for G', if $a \geq 3$ the preceding argument is valid. If $a=2$, so that $m=p/2$, note that every point, save $v_{p/2}$, which is adjacent to v_o is in $N^*(A)$; there are $\delta(G')-1$ such points. Moreover, since G' is not complete, $k < (p/2)-1$ and thus v_o is adjacent to neither $v_{(p/2)+1}$ nor $v_{(p/2)-1}$. Since both of these points are adjacent to $v_{p/2}$ ($r \geq 1$) they are in $N^*(A)$, so $|N^*(A)| \geq \delta +1$, a contradiction.

The graph $\bar{G}$ is the circulant with jump sizes $r+1$, $r+2, \ldots, [p/2]$. Thus v_o is adjacent to v_j if and only if $r < j < p-r$. The argument given above for G' in the case $a=2$ is valid for $\bar{G}$, so we assume $a \geq 3$. Let b be the least positive integer with $bm > r$; we claim that $(b+1)m \leq p-r$ and hence that v_o is adjacent to every v_j with $bm < j < (b+1)m$. In the case where $m \leq r$, note that $(b+1)m = (b-1)m+2m \leq 3r$ since $(b-1)m < r$. Since $3r \leq p-r$, the desired contradiction is attained. If $m > r$ then $p-m = (a-1)m < p-r$; since $a \geq 3$, we have justified our claim and thus have established that $\bar{G}$ has atomic number 1.

Now suppose p is even and consider $\bar{G}'$, which is the circulant with jump sizes $r+1, r+2, \ldots, (p/2)-1$. Since v_o is not adjacent to $v_{p/2}$, $a \neq 2$. Thus we assume $a \geq 3$. With b defined to be the least positive integer with $r \leq bm$, the argument of the previous paragraph establish that every point

v_j with $bm < j < (b+1)m$ is adjacent to v_0 unless $j=p/2$. If $a \geq 4$, so that $m \leq p/4$, the point $v_{(a-1)m}$ of A is adjacent to $v_{p/2}$ in $\overline{G'}$, since $(a-1)m-p/2 = p/2-m$ and $r < p/4 \leq p/2-m < p/2$. The proof is thus complete in this case. If $a=3$ then $m=p/3$ and thus the point of A are $v_0, v_{p/3}$, and $v_{2p/3}$. Moreover, by the above, the only point not known to be in $N^*(A)$ are their antipodes $v_{p/6}, v_{p/2}$, and $v_{5/p6}$ of the points adjacent to v_0 in $\overline{G'}$ only $v_{p/3}$ and $v_{2p/3}$ are not in $N^*(A)$. Since $p \geq 4r+3$ and $r \geq 1$, $p \geq 7$ and thus v_1, v_{p-1} are not in A and are also not one of the antipodes $v_{p/6}, v_{5p/6}$ of points in A. Thus there are at least two points of $N^*(A)$ not adjacent to v_0 and hence $|N^*(A)| \geq \delta$, contradicting our assumption that $a=3$. Thus $\overline{G'}$ has atomic number 1 and the proof of proposition 2 is complete.

4. SUFFICIENCY

The following lemma and its corollary are the final ingredients in proving the sufficiency for δ-realizability of conditions (1), (2), and (3) of our main theorem.

<u>Lemma 2</u> Suppose $p \geq 4$ and let G be a p-point graph with $2 < \kappa(G) = \delta(G)$ and $\kappa(\overline{G}) = \delta(\overline{G})$. Then there is a line of G such that $\kappa(G-e) = \delta(G-e) = \kappa(G)-1$ and $\kappa(\overline{G-e}) = \delta(\overline{G-e}) = \delta(G)$.
Proof. Let u be a point having minimum degree in G, and let w be a point having minimum degree in $\overline{G}$ with $w \neq u$. That such a point w exists if u does not have minimum degree in $\overline{G}$ is clear; if u does have minimum degree in $\overline{G}$, then $\delta(G) = \Delta(G)$ and hence G and $\overline{G}$ are regular graphs, and any point distinct from u will do. Since $\delta(G) \geq 2$ we may choose a line e of G incident with u but not incident with w. It then follows that $\delta(G-e) = \delta(G)-1$ and

u is a point of minimum degree in G-e; the degree in $\bar{G}$ of u is unchanged by the addition of e to $\bar{G}$ and hence $\delta(\overline{G-e}) = \delta(\bar{G})$. Morever $\kappa(\bar{G}+e) \geq \kappa(\bar{G}) = \delta(\bar{G}) = \delta(\overline{G-e})$ and thus $\kappa(\overline{G-e}) = \kappa(G)$.

Clearly $(G-e) \leq \delta(G-e) = \delta(G)-1 = \kappa(G)-1$. Thus let W be a minimum disconnecting set of G-e. Clearly u is not in W since G-e-u=G-u and $|W| < \kappa(G)$. If u is isolated in (G-e)-W, then $\kappa(G-e) = |W| = \delta(G-e) = \kappa(G)-1$ and the result follows. If u is not isolated in (G-e)-W, then (G-e)-W-u is still disconnected. Since G-e - w = G-w, W∪{u} is a disconnecting set for G so that $\kappa(G-e)+1 = |W \cup \{u\}| \geq \kappa(G)$, and the proof is complete.

Corollary If $(p,k,\bar{k}) > (3,0,0)$ is δ-realizable and $2 < k < \bar{k}$, then $(p,k-1,\bar{k})$ and $(p,k,\bar{k}-1)$ are also δ-realizable.

We are now ready to prove the sufficiency of conditions (1), (2), and (3), for δ-realizability. That (p,o, p-1) is δ-realizable may be seen by considering $G=K_p$. Thus, by repeated application of the corollary to lemma 2, every triple $(p,o,\bar{k})$ with $1 < \bar{k} < p-1$ is δ-realizable. If $(p,1,\bar{k})$ satisfies conditions (1), (2), and (3), then $\bar{k} < p-3$. The triple (p,2,p-3) is seen to be δ-realizable by considering the p-point circulant with jump size 1 (i.e., C_p); by the corollary to lemma 2, (p,1, p-3) is δ-realizable, and thus repeated application of that corollary implies that $(p,1,\bar{k})$ is realizable.

All that remains is to consider the triples $(p,k,\bar{k})$ satifying the conditions, where $2 \leq k \leq \bar{k} \leq p-3$. If k is even, say k=2r, then the p-point circulant G with jump sizes $1,2,\ldots,r$ satisfies $\kappa(G)=\delta(G)=k$ and $\kappa(\bar{G}) = \delta(\bar{G})=p-1-k$, and hence (p,k,p-1-k) is δ-realizable. It then follows by repeated applications of the corollary that $(p,k,\bar{k})$ is δ-realizable. If k is odd, say $k=2r+1$, and p is even, then the p-point circulant G' with jump sizes $1,2,\ldots,r$, $p/2$ satisfies $\kappa(G')= \delta(G')=k$ and $\kappa(\overline{G'})= \delta(\overline{G'})= p-1-k$. Thus $(p,k,p-1-k)$ is δ-realizable and hence the same holds for $(p,k,\bar{k})$. If k is odd and p is odd, then $k<p-1-k$. By the above the triple $(p,k+1,\bar{k})$ is δ-realizable and hence so is $(p,k,\bar{k})$, and the proof of the main theorem is complete.

We conclude by observing that, due to the sufficiency of our conditions for δ-realizability, the conditions are also necessary and sufficient for the existence of a p-point graph G with line-connectivity k and with $\overline{G}$ having line connectivity $\bar{k}$.

REFERENCES

[1] F. Boesch and A. Felzer, A general class of invulnerable graphs, Networks 2 (1972) 261-0283.

[2] F. Boesch and C. Suffel, Realizability of p-point graphs with prescribed minimum degree, maximum degree, and line connectivity, J. Graph Theory (1980) 363-370.

[3] F. Boesch and C. Suffel, Realizability of p-point graphs with prescribed minimum degree, maximum degree, and point connectivity, Discrete Appl. Math. 3 (1981) 9-18.

[4] F. Harary, Graph Theory, Addison Wesley, Reading, Mass (1969).

[5] M.E. Watkins, Connectivity of transitive graphs, J. Combinatorial Theory 8 (1970) 23-29.

ANNALS OF DISCRETE MATHEMATICS

Vol. 1: Studies in Integer Programming
edited by P. L. HAMMER, E. L. JOHNSON, B. H. KORTE *and* G. L. NEMHAUSER
1977 viii + 562 pages

Vol. 2: Algorithmic Aspects of Combinatorics
edited by B. ALSPACH, P. HELL *and* D. J. MILLER
1978 out of print

Vol. 3: Advances in Graph Theory
edited by B. BOLLOBÁS
1978 viii + 296 pages

Vol. 4: Discrete Optimization, Part I
edited by P. L. HAMMER, E.L. JOHNSON *and* B. KORTE
1979 xii + 300 pages

Vol. 5: Discrete Optimization, Part II
edited by P. L. HAMMER, E.L. JOHNSON *and* B. KORTE
1979 vi + 454 pages

Vol. 6: Combinatorial Mathematics, Optimal Designs and their Applications
edited by J. SRIVASTAVA
1980 viii + 392 pages

Vol. 7: Topics on Steiner Systems
edited by C. C. LINDNER *and* A. ROSA
1980 x + 350 pages

Vol. 8: Combinatorics 79, Part I
edited by M. DEZA *and* I. G. ROSENBERG
1980 xxii + 310 pages

Vol. 9:   Combinatorics 79, Part II
         edited by M. DEZA *and* I. G. ROSENBERG
         1980 viii + 310 pages

Vol. 10: Linear and Combinatorial Optimization in Ordered
         Algebraic Structures
         edited by U. ZIMMERMANN
         1981 x + 380 pages

Vol. 11: Studies on Graphs and Discrete Programming
         edited by P. HANSEN
         1981 viii + 396 pages

Vol. 12: Theory and Practice of Combinatorics
         edited by A. ROSA, G. SABIDUSI *and* J. TURGEON
         1982 x + 266 pages